The French
Marble Clock

The French Marble Clock

A Guide for Buyers, Collectors
and Restorers with Hints on Dating
and a list of Makers

Nicolas M. Thorpe

With a foreword by Michael Turner,
Department of Clocks and Watches,
Sotheby's of London

N.A.G. Press

First published in 1990
Reprinted 1993
Reprinted 1998
Reprinted 2002

Robert Hale Limited
Clerkenwell House
Clerkenwell Green
London EC1R 0HT

British Library Cataloguing in Publication Data

Thorpe, Nicolas M.
    The French marble clock: a guide for buyers, collectors and
    restorers with hints on dating and a list of makers.
    1. French marble clocks – Collectors' guides
    1. Title
    739'.3744

ISBN 0 7198 0230 X

Printed by St Edmundsbury Press Limited and bound
by Woolnough Bookbinding Limited

Contents

Acknowledgements

This book would not have been started, let alone finished, without the constant support of my wife, Valerie. I am particularly grateful for the encouragement she gave me during the difficult times that inevitably arose.

My sincere thanks go to Michael Turner of Sotheby's for writing the foreword and for reading parts of the manuscript. Michael was one of the first people to hear of my plans for a book and has always been enthusiastic about the project. I am also grateful to Gary Bateman, who voluntarily took on the task of reading the original manuscript.

I would like to acknowledge the help of Dr Robert Richardson, Anna and David Wingate, Martin Taylor, Catharina Lionetti-Maguire, Charles Allix, Chris H. Bailey and René Lecourtois. There are many others who must sadly remain anonymous.

My mother Marie, and my late father Charles, unwittingly laid the foundations for the book by providing additional tuition when I struggled with French as a schoolboy. The research would have been nigh-on impossible without the ability to write, read, and occasionally be understood in, French.

Nicolas M. Thorpe
1989.

Acknowledgements for illustrations: All photographs were taken by the author, unless indicated. The older illustrations in Chapters 1, 2, 11 and 12 were reproduced from period postcards and archival material.

Foreword

Originally, I became aware of the lack of written
information on French marble clocks in 1958 when, at
the age of eleven, I bought my first black marble clock
for a shilling (5p). In 1985, nearly thirty years later,
Nicolas Thorpe wrote to me outlining his ideas for a
reference book on the collecting and restoring of 19th
century French marble clocks. I was delighted with the
idea, not only because he was evidently very
enthusiastic about the subject, but also because these
excellent clocks have been sadly neglected and
undervalued for many years.

Many of the marble clocks one encounters nowadays are
dull and grimy shadows of their former glory. My shilling
purchase was a plain little marble clock with a Japy
movement. Friends and relations found it hard to
understand why I was enamoured of such a humble
clock. To me, however, the quality shone through the
dust and tarnish. At that time I already owned two
wooden cased clocks; an American Ansonia and a
German mantel clock. Although they were both worth
more than the little marble clock, it was plainly obvious
that the movements were produced to a much lower
standard than their French counterpart.

To date, the greatest frustration for the collector of
French clocks from this period has been the lack of
information. Some manufacturers were discussed in
Charles Allix's and Peter Bonnert's excellent book called
'Carriage Clocks', although it was naturally outside the
scope of their book to cover the production of pendulum
movements and their cases.

Although the heyday of the French marble clock was
only about a hundred years ago, there is very little
tangible evidence of their production left to assist the
collector with dating and identification. It is to Nicolas
Thorpe's credit that he has researched the subject so
thoroughly, and discovered so much previously

unrecorded information, despite the devastating effect on France and Belgium of two world wars that destroyed so much vital evidence.

In the Department of Clocks and Watches at Sotheby's, I receive many letters asking for information about French marble clocks, confirming my belief that there has been a revival of interest in them. They are still, at the time of writing, ridiculously underpriced, and I hope that Nicolas Thorpe's clearly written book will persuade the collector to buy a dowdy specimen and restore it to its former glory.

It is still amazing to me how the whole appearance of a dingy clock is transformed by cleaning the brass bezel and polishing the marble with wax until it gleams. I sincerely hope that many more people will come to love the French marble clock as a result of this book.

Michael Turner

Department of Clocks and Watches,
Sotheby's, London.

The History of the Marble Clock part I

French marble clocks were sold in large quantities from 1850 until the beginning of the First World War in 1914. War and a change of fashion ended the production of one of the most popular clock types that has ever been produced. The Free Trade Treaty of 1860 between Great Britain and France allowed clocks, among other items, to be exported in large quantities free of any tariffs. This may explain why so many French marble clocks are to be found in Britain. These clocks were some of the first to be produced in large quantities and were fitted with the classic pendule de Paris movement (Fig 1.1).

This type of movement was developed in Paris towards the end of the eighteenth century and became one of the standard movements for French mantle clocks. In the eighteenth century the movements were made by a relatively small number of master craftsmen with each component being hand made; the production of a single movement was consequently an expensive and time consuming process.

Towards the end of the eighteenth century the demand for good quality movements for mantle clocks began to outstrip supply. At the beginning of the nineteenth century, entrepreneurs such as Frédéric Japy, took advantage of the situation and started to manufacture movements in large quantities using machinery and methods that closely resemble modern mass-production techniques.

Fig 1.1 The classic pendule de Paris movement that is found in many French marble clocks.

1

An Englishman, Julien Tripplin, writing under the heading 'Watch and Clockmaking at the Paris Exhibition of 1889' noted:

"Paris is known all over the world as the great emporium of the clock trade. Nevertheless the clocks actually made there are few in number, and comprise only the very best astronomical regulators made by hand and at a high price. But the various movements of St. Nicolas and Beaucourt districts are the cause of about 8,000 hands being employed in that city, directly or indirectly, in the clockmaking and finishing trade, which subsequently gives work to... marble cutters and polishers."

The movements were either called 'blancs', consisting of a frame made of the plates, pillars and barrels, or 'blanc roulants' which were the same but had the pivoted train wheels present. The movement finishers in Paris added the escapement and other working parts of the movement, and the quality of the overall movement was maintained by the stringent standards set by the Parisian clockmakers.

Companies such as Japy originally produced movements for a wide range of Parisian mantle clocks. Later, in 1882, they started to produce complete marble clocks using their own movements and cases. The Japy catalogue of 1898 includes forty pages of marble clocks. Similarly the companies in Saint-Nicolas d'Ailermont produced complete clocks.

Many of the pendule de Paris movements found their way into marble cases that were produced in England and the U.S.A. The Americans also fitted their own movements into cases made of cast iron that were enamelled to look like black marble. Tin sheet was lacquered to give it the appearance of onyx and other decorative marbles. Britain used its own deposits of good quality black marble to make cases, although these were not produced in any significant volume. Chapter Ten on clock cases gives more information.

The Movement Makers

Most French marble clock movements originated from two areas in France: Saint-Nicolas-d'Aliermont in the north, and Montbeliard in the east. In the Montbeliard region were the workshops and factories of Japy, Vincenti, and Marti. We are fortunate that a lot of

2

historical material is available for Japy, but sadly little is known about Vincenti and Marti.

Jean Vincenti & Roux

Jean Vincenti, sometimes known as Vincent, was of Corsican origin and established himself in Montbeliard in 1791. In 1823 a factory was opened using machinery that he had designed for the production of rough clock and watch movements.

Roux is said have taken over the company in 1824 when Vincenti went bankrupt, but continued to use the Vincenti name. By 1825 it was reported that the factory employed 30 to 40 workers. Production is said to have ceased in 1870 and sadly, little more of the company is known. Awards to 'Vincenti' were:-

Médaille d'Argent 1834
Médaille d'Argent 1855

Japy, Marti & Roux

Around 1864, three separate companies joined to form Japy, Marti & Roux, and represent their products in Paris. In 1867 they were in Boulevard du Prince-Eugene and 1880 Rue Turenne. (Tardy [5].)

Samuel Marti et Cie

The name Marti is frequently found on the back plate of French marble clocks movements. Once again little is known of the company. Two companies are known by this name, Samuel Marti et Cie. of Montbeliard, and Fritz Marti of Vieux Charmont.

S. Marti are reputed to have produced pendule de Paris movements between the years 1841 and 1912, although there is little formal documentation to substantiate these dates.

S. Marti et Cie. exhibited in Paris during the years 1841, 1851, 1852. The following awards are known to have been made:-

1860 - Samuel Marti, Médaille de Bronze.
1900 - Samuel Marti, Médaille d'Or.

5 Tardy.
Dictionnaire des Horlogers Français. Paris 1971. The standard reference work for anyone wishing to date French clocks. Unfortunately it is very expensive although only available in paperbinding.

3

The company had premises in Rue Orleans-Marais in 1850 and Rue Charlot in 1860 (Tardy [5].)

Fritz Marti

The name of Fritz Marti is occasionally encountered on the movements of French marble clocks. Unfortunately little is known except that there were premises in Vieux-Charmont in 1876.

1900 - F. Marti et Fils Médaille d'Or, Paris.
Médaille de bronze, (date not known).

The Japy Company

The Japy Company was a phenomenon of nineteenth century France, producing clock and watch movements in quantities that were large even by modern standards. It was the first company in France to mass-produce unfinished watch and clock movements. Mass production did not, however, lower the quality of the finished products. Japy clock movements had the reputation of being second to none and demand for the fine products often outstripped supply.

Between 1770 and 1888, the company produced 3.5 million watch movements and during 1888 alone, they supplied 1,200,000 of these movements to the casemakers of Switzerland and Besançon. In 1889 the Badeval and Beaucourt factories were reputed to employ some two thousand people with large machinery having a quoted rating of 800 horsepower. It should be remembered, however, that not all of these people were involved with horology. Between 1810 and 1888 the company manufactured over 6 million blank clock movements; many of these were sent to Parisian clockmakers. In 1889, some 500,000 movements were produced with a value of £120,000.

5 Tardy.
Dictionnaire des Horlogers Français. Paris 1971. The standard reference work for anyone wishing to date French clocks. Unfortunately it is very expensive although only available in paperbinding.

Fig 1.2 Frédéric Japy (1749-1812), the founder of the massive Japy company. In 1985 the bust stood in the derelict grounds of a demolished Japy mansion in Beaucourt. (Photo: 1985).

Frédéric Japy

The founder of the Japy company was Frédéric Japy (Fig 1.2), an enterprising man, born in Beaucourt, France, on May 22nd 1749. Beaucourt is located near the Swiss

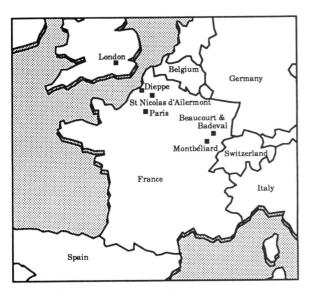

Fig 1.3 The two main regions of marble clock movement production in France – St Nicolas d'Ailermont and the Montbéliard region.

border, as shown on the map in Fig 1.3. He was the second son of a family of nine children, and his father Jacques was a blacksmith and cartwright in Beaucourt. The smithying work was undertaken in a yard in front of the Japy home, while at the back was a small workshop where the father ran a locksmith's business. Jacques Japy was an influential character in Beaucourt and in 1760 he was made Mayor by the Prince of Montbéliard-Wurtemberg.

It is worth noting that Frédéric Japy's father was a locksmith as well as a blacksmith; by modern standards these occupations are considered to be fairly basic, but in their time they were among the most skilled occupations that a person could undertake. Frédéric Japy had a good start in life, inheriting technical prowess from his father coupled with a shrewd yet compassionate business ability.

At the age of seventeen he left school and stayed in Beaucourt for a while before going to Le Locle in Switzerland for three years as a watchmaking apprentice. After several months he had made his mark as a gifted worker and his patron predicted a brilliant future for him.

Japy's First Horological Workshop

Frédéric returned to Beaucourt in 1772 and immediately set up a watchmaking workshop, apprenticing many of the more able young people of the district. The small company produced ebauches; these were unfinished watch movements usually consisting of the plates, bridges and cocks. The finisher of the watch would often fit the mainspring and jewels and maybe produce the escapement.

In common usage, the French word 'ebauche' means a rough sketch, and an 'ebauchoir' is a sculptor's chisel that is used for roughing out large blocks of stone. The making of watch ebauches was traditionally performed by hand, a long and time consuming process. Japy used his mechanical genius to automate some of the

5

operations, aided by the talented mechanical engineer, Jeanneret-Gris, from Le Locle.

The first Japy factory was destroyed by fire, started by a troop of Austrian soldiers in 1815. It was immediately rebuilt to form a larger concern that easily accommodated new machinery and an ever increasing workforce. The building, parts of which date from 1803, is shown in Fig 1.4.

Japy was a kindly man and became a patriarch to his workforce. His consideration helped the company to survive many of the political and social upheavals that plagued French history during the nineteenth century. In the early days, the staff lived with the family and in later years the company was one of the first to provide medical care for its workers; the Japy philanthropic ideals were later extended to financing the erection of houses that were sold to employees against loans provided by the company.

In anticipation of further expansion, the water mill at Badeval was purchased in the year 1800 together with the mills at Lafeschotte, d'Etupes, de Berne, Seloncourt, and Vandoncourt. He also bought the convent of Bellelay, the source of the factory bells for Badeval and Beaucourt.

Badeval and the Production of Movements

In the early part of the nineteenth century clock movements were usually built by hand, a slow and expensive operation. Japy decided, with typical flair, to mechanise clockmaking using the same methods he had applied to watchmaking. Badeval, a distance of some 3 kilometres from Beaucourt, was chosen to be the factory for clock movement production.

The old water mill was powered by a spring forming the water source, La Fechotte. The existing mill building was extended and improved to provide a clock factory that stood from 1810 until 1979, when it was demolished by the current owners of the site. Fig 1.5 shows the original mill before its demolition.

The machinery used in Beaucourt for making watch ebauches was quickly adapted to meet the requirements of clock movement production. For this reason, work on the new factory proceeded rapidly and, by 1810, the

Fig 1.4 The main Japy factory in Beaucourt, parts of which date from 1812. This is adjacent to the original family workshop. (Photo: 1949).

Fig 1.5 The original Japy factory in Badeval. It was originally a watermill. (Photo: circa 1978).

factory was in full production. The enterprise was an immediate success and the Japy movements were bought in Paris as quickly as they were made.

In 1806 Japy's five sons formally joined the company. They were Frédéric Guillame (better known as Fritz), Louis, Pierre, Frédéric (called Fidot), and Charles. The first three looked after Beaucourt while the two youngest, Charles and Fidot, managed the watch and clockmaking activities at Badeval.

Death of Frédéric Japy

Fig 1.6 A monument to Charles Japy, one of the Japy brothers. In 1985, it stood in a small and neatly kept vegetable garden opposite the main Badeval factory buildings. (Photo: 1985).

In 1811, with business booming and continued prosperity apparently assured, Frédéric Japy's wife Catherine Suzanne died. Frédéric died soon after, on June 23, 1812, at the age of 63; it has been suggested that he died of a broken heart. The Japy estate was subsequently divided among the five sons, with the daughters receiving money for their shares of the company.

Charles was to die in 1821 at the age of 28, leaving Fidot to continue the work at Badeval alone. A critical period followed at Badeval, that did not end until 1828 when the three Japy brothers at Beaucourt bought the factory from Fidot. A monument was erected to Charles, presumably by Fidot, in 1821. It now stands rather forlornly outside the Badeval factory gates in a small triangle of land that is now a vegetable garden (Fig 1.6).

The inscription is strangely worded and, in translation, reads:

"Very near this monument which was erected by the one that cared, under these young poplars out of sight of the wicked ones, rests, alone the body of one of the best children. If he did not reach old age, he lived long enough to know that here on earth there are cheats hiding around every corner. Jean-Charles Japy born 17 January 1792 died the 25th May 1821".

8

The strong impression is that Charles and Fidot were in disagreement with the Japy brothers who were eventually to rescue Badeval from commercial difficulties. Jean-Charles died in May 1821, three months after the brothers in Beaucourt had formed a separate company.

At this stage, Ingenu, son of Fritz Japy, became manager of Badeval. Ingenu had been a pupil of the celebrated horologists Sadoz and Breguet. He made radical changes to the production of clock movements, inventing, among other items, machines that automatically cut pinions. Under his able direction, Badeval became the premier factory in France.

Formation of Japy Frères

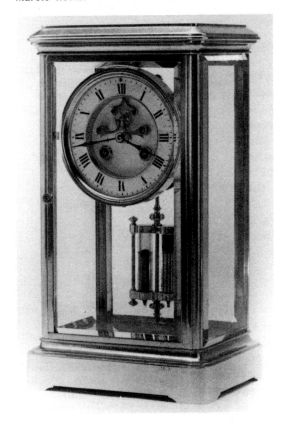

Fig 1.7 A Regulateur de Paris. The Japy movements in these popular clocks are identical to those used in French marble clocks.

On February 3rd 1821, Fritz, Louis, and Pierre Japy formed the first company under the name of 'Japy Frères'; this was registered again twenty-five years later, on August 1st 1837. It would appear that company names could only be registered for a relatively short period of time. The law was changed by the time of the last registration in 1928, which will last until the year 2000.

In 1854 the old company of Japy Frères was dissolved and replaced, on August 1st 1855, by Société Japy Frères et Cie. This name lasted for twenty years until 1875.

In 1863, or possibly 1865, a company was formed to represent the interests of three local horology manufacturers, namely Japy, Marti, and Roux, with shops in Paris. (See Appendix for details of Marti and Roux.)

The three Japy brothers responsible for Beaucourt died within eleven years of one another; Louis in 1852, Fritz in 1854, and finally Pierre in 1863.

Japy started to fit their own watch movements into cases in 1850. Before this they had always supplied movements to other watchmakers. With the development of some clever machinery, the company was able to start mass

9

Fig 1.9 The movement from
the Japy clock illustrated
in Fig 1.8. This is an inferior
product sometimes called a
'tic-tac' movement.

producing watch cases in a wide range of metals and finishes.

Later, in 1860, the same step was taken with clocks, initially to produce wooden cased clocks and 'pendules tole', which are believed to have been made out of sheet metal that was painted and decorated with a wide variety of scenes and designs. They were regarded as kitchen clocks and rapidly gained a reputation for their reliability. At the same time Japy produced the first regulateurs of the type shown in Fig 1.7. These clocks used the classic pendule de Paris movement. The same type of movement was fitted into French marble clocks.

Japy Marble Clocks

In 1882, the Japy company developed a process of gilding called 'la lustrerie'. The main objective was to profit from the great demand for the type of French clock that was made of gilded bronze or ormolu. The process originated in the seventeenth century and was used to decorate the cases of some

Fig 1.8 A Japy marble clock of
rather poor quality. This was
found in 1986 in a French shop
on the Normandy coast.

N° 3070. — PENDULE MARBRE GRAVÉ
NOIR ET COULEUR, ORNEMENTS BRONZE
Hauteur 42 cent. — Largeur 25 cent.

N° 4017. PENDULE LANTARA
MARBRE ET BRONZE IMITATION
Hauteur 44 cent. — Largeur 27 cent.

Fig 1.10 One of forty pages of French marble clocks from the Japy catalogue of 1898. The top clock is similar to that in fig 1.8.

84 *Vinter.*
Histoire des Establissments Japy Frères (1777 - 1943). Japy Frères, Beaucourt, France. 1944.

French clocks. Later complete cases were made of ormolu. Japy designed and installed a bronze casting and gilding plant. This new process was the most successful that the company had ever introduced.

A cheaper metal made of zinc (spelter) was used to produce much cheaper cases. At the same time, the spelter process was used to produce marble mantlepiece ornaments. From the original work on clocks, the industry was extended to include 'articles deluxe' such as jewel boxes, powder puffs, electric lamps and a formidable array of other products.

Large saw benches were installed in the factory at Lafeschotte to process timber for all the Japy factories. The host of wooden goods even included packing cases. The same plant was probably extended to incorporate the cutting of marble for the lustrerie business and eventually for the making of marble clock cases. Vinter[84] specifically states that in 1882 Japy was producing, marble cases and candlesticks with ormolu embellishments.

Large Marble Clock Range

I found a Japy marble clock in a French junk shop in 1986. It is shown in Fig 1.8. The movement (Fig 1.9), is not the pendule de Paris type normally associated with a French marble clock, but an inferior one, sometimes called a 'tic-tac'. The clock appeared in Japy's 1898 catalogue on the page reproduced in Fig 1.10, one of forty pages dedicated solely to marble clocks. Fig 1.11 shows the cover.

11

End of the Japy Empire

The period around 1898 appears to have been the heyday of the Japy company. This year saw the formation of their electrical products division, one that grew to produce electric motors, switch-gear and a host of industrial products using electricity. In 1899 the company started to make photographic equipment, followed by gramophones in 1900 and typewriters in 1910.

The watch and clockmaking industry went into a decline after the First World War (1914-1918) and the Badeval factory finally stopped producing clocks in 1936. The watchmaking side of Japy had been liquidated in 1934 after a major rationalisation of the whole company in 1933. The end had come after 164 years of producing top quality horological goods that were considered by many to be supreme in their class.

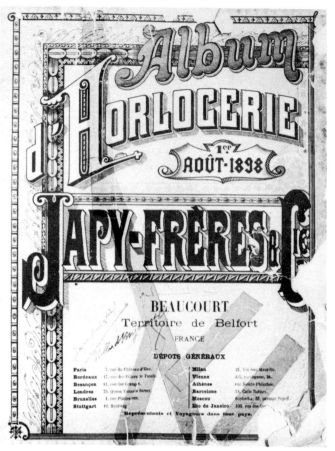

Fig 1.11 The splendid cover of Japy's 1898 clock catalogue.

During this period the company won many awards at trade fairs and international exhibitions. Here is the citation from the Exposition Universelle de 1855 that was held in Paris and was equivalent to one of the English Great Exhibitions:

"Class 8: Japy Brothers of Beaucourt (France). Large scale mechanised manufacture of watch ebauches and blanc roulants for clocks. Good quality and low price".

In the section Grand Jury Awards - Exposition Universelle de 1855, Paris, Japy and a map-making company were the only winners of the award, an indication that the awards were not easily obtained. Listed here are the awards won by Japy between 1819 and 1880. Refer to Makers and their Marks at the end of this book for further information on dating.

12

All the awards were for clock or watchmaking; items 1867 to 1879 are believed also to be for horology, but this cannot be verified.

Gold Medals:	Louis XVIII	1819
	ditto	1823
	Charles X	1827
	Louis Philippe	1834
	ditto	1839
	ditto	1844
	ditto	1849
Grand Medal	Great Exhibition,London	1851
1st Class Medal	New York Exhibition	1853
Grand Medal of Honour	Exposition Universelle, Paris	1855
ditto	ditto	1867
Grand Diploma of Merit	Exposition Universelle, Vienna	1873
Grand Prix d'Honneur	Exposition Universelle, Paris	1879
Grande Diplôme d'Honneur	Exposition Regionale, Besançon	1879
ditto	Exposition speciale d'horlogerie de la Chaux-de-Fond	1880

Fit 1.12 The Japy building in Badeval where pendule de Paris clock movements where produced. It was built in 1888 and in 1985 was awaiting demolition. The klaxon on the roof was driven by electricity, replacing a bell that came from a nearby convent. (Photo: 1985).

The Japy Factory in Badeval in 1985

Driving through Badeval on a hot July day in 1985, it was difficult for me to equate this sleepy little village to the enormous volume of horological work that had emanated from it. At first I wondered whether I had come to the right place, but was eventually rewarded by the sight of the main building standing on a bend in the Grande Rue (Fig 1.12). In front of the buildings stood a small triangle of ground that had a laundry pool, and a terrace of old houses dating back to the early days of the Japy factory. The other side of the triangle led to a small vegetable garden where the monument to Jean-Charles Japy (1729-1821) stood (Fig 1.6).

In 1888, the building shown in Fig 1.12 was where pendules de Paris movements were produced for marble

clocks and for the familiar French four glass regulateurs de Paris. The entire premises were bought in 1972 by a French light engineering company called Etablissments Peuquet; although no longer connected with Charles Pequet, the founder, the company retains his name. Pequet built his first workshop in 1871 at Seloncourt, a town lying some 10 km to the south-west of Badeval.

In Badeval I met the company's commercial manager, Christiane Hintzy, who showed me the parts of the original Japy buildings and indicated the changes that had been made to the original site. The part formed by the corn mill was demolished in 1979. A large iron staircase had fallen down at some time during the eight years before demolition, as is obvious from Fig 1.5, where the state of the building can be seen shortly before its demolition; the staircase lay on the ground and the whole structure was declared unsafe to enter. Pequet later demolished more buildings, leaving only the main block and the newer single storey units lying to the left of it.

The main block (Fig 1.12), where the pendules de Paris movements were manufactured, has three floors, but by 1985 the two upper floors had become unsafe, while the ground floor was being used as a store for lubricating oils for the Pequet machinery. Originally the ground floor housed heavy machinery. Assembly work and lighter operations were carried out on the two upper floors.

In Fig 1.13 showing the end of the main block, the two upper windows were used as offices, while the little structure and window to the right provided

14

accommodation for a gatekeeper. The large three-trumpeted klaxon mounted on the delicate roofing was totally incongruous in this quiet little French country village. It was driven by electricity and had replaced the bell used as the works alarm before the days of electricity.

Plans for future development of the site seemed rather vague because someone was still living in the gatehouse. While taking photographs, I was approached by Marc Reiniche, owner and occupier of the house he had bought in 1965, shown on the left in Fig 1.14. The Badeval factory manager had once lived there. M. Reinche had worked for Japy in Beaucourt between 1947 and 1962 and said his father, Emile Reiniche, was alive and living in the gatehouse.

M. Emile Reiniche had joined Japy in 1936 as 'gardien' or gatekeeper, but by 1940 all work had ceased at Badeval. In 1941 he went to Beaucourt to undertake the same work for Japy until he retired in 1965 at the age of sixty-five. The Badeval premises were used after 1940 as a store for the products of the nearby Peugeot factory; a local newspaper article dated 1955 states that velomoteurs (mopeds) were stored there.

The Reiniches provided some invaluable information, including names of people who were known to have worked for Japy, photographs and other documents. Through M. Reiniche I met André Brand, an old colleague of Emile Reiniche, who also remembered the names of people who had worked for Japy at the end of the nineteenth century.

He directed me to Maurice Elbling, who lived in a large house just outside Beaucourt. He ran a small clock and watch repairing business from a room on the top floor with his son André; they were in fact horologers. Maurice Elbling had worked in the Japy watch factory in Beaucourt until it closed in 1934. The small neat workshop with benches set

Fig 1.14 A row of Japy company houses in Badeval; the house on the left was originally the Badeval factory manager's residence. (Photo: 1985).

against two large windows overlooking the street below, gave the impression of belonging to an era when many hundreds of local people were employed in their own households as outworkers for the clock and watchmaking districts of France and Switzerland.

M. Elbling stated that pendule de Paris movements were only ever made in Badeval. They were assembled in the building where M. Reiniche now lived, a fact that was corroborated by Marc Reiniche himself. Unfortunately a mystery still remains about the building because M. Reiniche indicated that it was erected in 1888, but Vinter's history of Japy suggests that 1921 was the date. My belief is that it dates from 1888 and that its pitched roof was added in 1921. An old postcard shows the original flat-topped building. It would be reasonable to suggest that the building in Fig 1.13, was used for the manufacture of pendule de Paris movements from 1888 onwards. Before that date they were probably produced in the older building shown in Fig 1.5.

Beaucourt and Badeval are well worth a visit because a Japy Museum was opened in 1986 on the site of the original Beaucourt building shown in Fig 1.4. In the countryside around Beaucourt are many derelict Japy factories and the people of the area are normally pleased to talk about a company that had such a remarkable influence over the area for nearly 150 years.

II

The History of the Marble Clock - part II

Saint-Nicolas-d'Aliermont (Fig 2.1) has been described as 'the cradle of (French) horology'. This large village provided many thousands of movements for French marble clocks and ranked in equal importance with the eastern part of France, where Japy Frères, Marti and Vincenti worked.

The Birth of Horology in Saint-Nicolas D'Aliermont

Fig 2.1 General view of Saint - Nicolas - d'Aliermont in the nineteenth century.

Following the death of their father in 1725, Charles-Antoine Croutte and his sister, Marie-Anne, took the forest road from Arques and headed for the nearby village of Saint-Nicolas d'Aliermont. Their father, Antoine Croutte, had been a clockmaker in Arques since 1702. Before this he was a clockmaker in Dieppe. His reasons for settling in the more peaceful countryside are

unknown; we do know, however, that the English bombarded Dieppe from the sea in 1694.

Saint-Nicolas was already the home of skilled metalworkers who had an abundant supply of fuel from the surrounding forest.

Charles-Antoine Croutte set up a small workshop in the village, unwittingly laying the foundations for one of the largest clock and watchmaking communities that would ever be seen in France. On October 17, 1727, he married Anne-Marie Legrand, who gave birth to twelve children including two boys: Charles in 1737 and Nicolas in 1739. Seventeen grandchildren followed to form a dynasty of horologers that was to be active in Saint-Nicolas for more than a hundred and fifty years.

Numerous young people were keen to learn the trade and many family names appeared, that later became celebrated names in French horology, such as Jean-Baptiste Legrand, Jean-Baptiste Delasbos, Jean Dumouchel, and Nicolas and Guillaime Papin. Free of the inherently restrictive practices of a guild system, the workers were able to develop their ideas as they wished.

A weight-driven clock movement was eventually produced that was peculiar to Saint-Nicolas. Under the guidance of Nicolas Breton, a sculptor and engraver, a distinctive style of clock also evolved; this was usually decorated with wood carved in the shapes of flowers and fruits.

Saint-Nicolas continued to prosper until the Revolution of 1789 created economic problems that left it in a state of considerable poverty. There was an acute shortage of work in the traditional trades of lacemaking and horology. One writer noted in 1790, that 645 workers from these two industries were destitute, and by 1791, some workers had only worked for ten days out of the year. The clock vendors of the village took their wares further afield in an attempt to secure some trade. At this time the principal wholesalers were Cauchy, Cailly, Quetteville, Hollingue, Bellengreville, and Dumouchel.

Honoré Pons

By 1807 the plight of Saint-Nicolas had become sufficently bad for a government official to take action. M. Savoye-Rollin, Prefect of La Seine-Inferieure, of which Saint-Nicolas was a part, sent a detailed memorandum to M. de Champigny, Minister for the Interior and also Minister for Clocks and Watches.

The memorandum described the precarious nature of the horological industry of Saint-Nicolas and asked for the services of someone who was an expert in the latest horological techniques. The Minister immediately sent Honoré Pons, a young Master Horologist, also known as Pons de Paul. Pons was the son of an instrument maker in Grenoble, he had learnt his craft in Paris and was credited with being familiar with the work of the celebrated maker Lepaute. He was a fine horologist and an accomplished mechanical engineer.

On arrival in Saint-Nicolas, Pons met Matthieu Croutte, a descendant of the Croutte family mentioned earlier. Croutte had already devised new methods of making movements and had also undertaken extensive research into the technical aspects of clock and watchmaking. Pons formed a company that was simply called 'Pons'. He was helped in this venture by Croutte and an Italian, Paul Rosetti.

New machines were evolved, capable of cutting, rounding, and polishing. Pons also split the work force into separate units, each producing one section of the components of a complete movement. It was the same division of labour that had proved very successful for Frédéric Japy in Beaucourt.

Although Pons managed, within two years, to achieve a considerable reduction in the cost of movements, the work was still mainly performed by hand. By comparison, Frédéric Japy in Eastern France was producing machined watch movements and was planning the mechanised Badeval clock factory that would open in 1810.

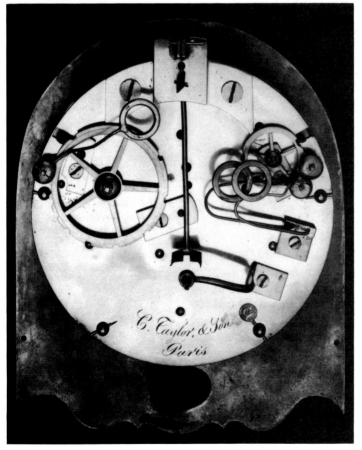

Fig 2.2 A large and unusual quarter striking movement bearing the Pons Médaille d'Or 1827 stamp adjacent to the lower right pillar. (Not from a marble clock). Photo: Sothebys, London.

Pons received many awards for his work including a silver medal in 1823, and gold medals in 1827 and 1839. Much of the company's work went direct to the clockmakers of Paris. A fine example of a Pons movement is shown in Fig 2.2.

Pons, like Japy, appears to have been a benevolent man. In his Will, which was read on July 19th 1857, he left a legacy of 1,000 Francs that was intended to "help the workers of this manufacturer when old age or infirmity prohibits them from working". The kindly gesture eventually led to problems with its distribution that were only solved by litigation. Not everyone wished to work in the new workshops and many people continued in their traditional work places at home that were usually sited in an attic with large windows to provide as much daylight as possible (Fig 2.3).

Fig 2.3 *Rue des Horlogers in Saint-Nicolas-d'Aliermont, Many houses in the town were fitted with vitrines, (attic windows for the use of horologists). (Photo: 1986).*

Fabrique d'Horlogerie de Saint-Nicolas-d'Aliermont

Pons realised that the traditional way of working in Saint-Nicolas was not particularly efficient. When he arrived, each clockmaker would make a complete movement from start to finish. Pons encouraged the local people to work together, with each craftsman making one particular part of a clock, enabling him to perform his work more quickly and efficiently.

The work was co-ordinated by Pons under a form of horologists' guild called 'Fabrique d'Horlogerie de Saint-Nicolas-d'Aliermont'. At the Paris Exhibition of 1819 the guild was awarded two silver medals for clock movements 'brut et en blanc' (rough movements), although the movements were not finished to the same degree as later ones.

In the exhibition report, the jury awarding the medals cited Fabrique d'Horlogerie de Saint-Nicolas for 'careful work and sound principles'. One medal was awarded to Pons in the same report, with the comment: "The Jury would particularly note their satisfaction with which they have seen the products of the clock makers of Saint-Nicolas-d'Aliermont and award a silver medal to M. Pons, to be deposited at the town hall of Saint-Nicolas-d'Aliermont."

Augustin Croutte & Cie

Augustin Croutte was the grandson of Matthieu Croutte, who in 1847 wrote a paper, Etablissment d'une Fabrique d'Horlogerie a Saint-Nicolas-D'Aliermont. It describes a factory specifically set up to produce the type of movements used in French marble clocks. The document currently resides in the French archives and provides a detailed insight into the horological industry of Saint-Nicolas in 1847.

Croutte explained his reasons for building a factory, giving detailed costings and estimates of production and expenditure. In 1847 he estimated the total number of clock movements produced in France to be in the order of 100,000 a year, mainly by Japy at Badeval and Vincenti of Montbéliard. Of this total figure, Saint-Nicolas made a mere 14,000 a year. He suggested that Japy and Vincenti had been successful, not just through mechanisation, but also by having a depot or warehouse in Paris where movements could easily be obtained when required by clockmakers.

Japy and Vincenti, while being heavily mechanised, in Crouette's view did not have the same degree of expertise that lay within the workforce of Saint-Nicolas. This expertise lead, so we are told, to the Saint-Nicolas movements being produced to a high finish that was appreciated by many of the Parisian clockmakers who had grown tired of the output from Vincenti and Japy.

Croutte was inspired by the ingenious idea of using machines to do the bulk of the work and leaving the finishing of the movements to the skilled workers of Saint-Nicolas, who could undertake the work in their own homes - the traditional workshops of the area.

21

The company intended to produce 10,000 movements a year with calibres varying between 3 pouce and 4 pouce. (Sizes are explained in Chapter Four). From Croutte's notes about quantities, we gain an idea of the popularity of certain calibres in 1847:-

Calibre:	2 3/4 Pouce	2,000
	3	4,000
	3 1/4	2,000
	3 1/2	500
	3 3/4	500
	4	1,000
Total:		10,000

It can be seen that the 3 pouce calibre, also called 'l'industriel', was the most popular at the time. A work force of seventy-five men and twenty-five women was estimated to be suffcient to produce these 10,000 movements. In other terms, each worker would produce the equivalent of just under two movements a week. Of the 2,000 inhabitants of Saint-Nicolas in 1865, about 1500 were earning an income through horology, with 800 of them being directly involved in the production of carriage clocks, and pendule de Paris movements of the type fitted into marble clocks.

Fig 2.4 The house in Saint-Nicolas-d'Aliermont of the chronometer maker, Onesieme Dumas. It had been restored shortly before 1986. Note the clock in the middle of the roof. (Photo: 1986).

Chronometers

Saint-Nicolas has always been associated with top quality work that displayed a good knowledge not only of volume production techniques but also of the technical aspects of horology. The last half of the nineteenth century saw the production of chronometers that were a visible indication of the skill and ability of the town's technicians.

Jean-Aime Jacob started a workshop in 1841 to produce regulators, chronometers and astronomical regulators. Another famous horologist was Onesieme Dumas (1824-1889), whose house is shown in Fig 2.4.

22

The picture was taken in 1986 after the house had been carefully restored. Note the clock on the top at the front. The house was later occupied by another chronometer maker, Delepine.

Around this time a school of horology was built in Saint-Nicolas at Blesdal. The largest manufacturer in the town, M. Villon, is believed to have provided the main initiative. Unfortunately the school was poorly attended, some say because of the hostility shown towards its establishment by the Mayor of St. Nicolas at the time. It was governed by M. and Mme Delaune and its closure must have been a sad loss to a town that depended heavily on the technical abilities of its young people.

Fig 2.5 A small workshop standing in the grounds of a house in Saint-Nicolas d'Aliermont. (Photo: 1986).

Fig 2.6 The original Couaillet factory in Saint-Nicolas d'Aliermont. This was burnt down in 1912.

Couaillet Frères 1891-1925

The Couaillet brothers, Gustav, Armand, Henri, and Ernest, formed a company in 1891 to produce pendule de Paris movements. Later they became famous for carriage clocks. Armand Couaillet, who led the brothers in their enterprise, had trained with M. Villon's company. Armand Couaillet was born in 1865 and, at the age of 27,

Fig 2.7 The house and factory of Delepine Barrois that was acquired by Couaillet Frères after 1912. The glass door shown in fig 2.8 is behind the fourth man from the left.

Fig 2.8 A glass door that originally lead to the offices of the building in fig 2.7. It has been etched with the shape of a pendule de Paris clock. (Photo: 1986).

he moved into a small workshop and worked with his brothers and some other qualified horologists. He made carriage clocks for the Parisian market with a high quality finish that soon produced many orders.

By 1902 the company employed 150 workers and at one time produced pendule de Paris movements that inevitably would have found their way into marble cases. Fig 2.6 shows the original factory; it was burnt down in

Fig 2.9 A page from the Exacta catalogue that was published sometime after 1900.
The text on the right reads:
"The high interchangability of our parts greatly simplifies your repair work".

1912. After the fire, the company bought the house and factory of Delepine Barrois (Fig 2.7). The glass door in Fig 2.8, with a delicately etched pendule de Paris clock on it, was originally the entrance to the offices. The factory and house in 1986, were owned by a private family and the door led into the kitchen.

Around 1930, Couaillet Frères experienced financial problems that proved to be the end of the company as it was known. The three brothers split up and Armand Couaillet went to work for Exacta.

25

Exacta and Lebrejale

Exacta was the trade name of a company called Lebrejale, which was founded around 1900 and ceased working in 1935, apparently having been financed by a Swiss organisation. The factory was in Saint-Nicolas with an office in Rue de Franche Comte in Paris. It produced a 14-day striking pendule de Paris movement, illustrated in Fig 2.9 together with a list of spare parts that could be ordered for each movement. The illustration was taken from an undated Exacta catalogue that included an eight-day timepiece fitted with a platform escapement in contrast to the striking movement that had a pendulum.

The history section of the catalogue gives the impression that the company had been in existence since the days of Honoré Pons and had collected all the various gold and silver medals that were attributed to him or his efforts. Exacta claimed a workforce of 900 and a factory of 11,000 square metres. René Lecourtois of Saint-Nicolas has declared that the history was wishful thinking by Exacta.

Albert Villon

Julien Tripplin[81] in 1898 described the factory of M. Albert Villon as being the largest in Saint-Nicolas with an annual production of 20,000 carriage clocks, 20,000 mantle clock movements, and 60,000 alarms. Alarms of this period bore little resemblance to modern clockwork alarms.

Villon claimed that his system of manufacture was independent of any outside company. This in effect means that his workshops would have been able to produce the brass, dials, and a host of other highly specialised items that made up a clock. Even if this were not completely true, it does show the extent of industrialisation which was comparable to Japy Frères in the same period. The factory was built in 1867 and Villon ran it for twenty years until 1887.

Duverdy and Bloquel - Reveils Bayard

Paul Duverdy became the manager of Albert Villon's company in 1887 and held the post until 1910. On the

81 Tripplin.
Watch and Clockmaking at the Paris Exhibition of 1889: Being an Account and Comparison of the Exhibits in the Horological section of the French International Exhibition. London 1889.

Fig 2.10 Part of the Reveils Bayard factory in Saint-Nicolas d'Aliermont. It was originally run under the name of Duverdy et Bloquel. (Photo: 1986).

death of Villon, he was joined by Joseph Bloquel, who became manager until 1922. After the First World War, the company turned its attention towards the domestic market and created the familiar alarm clocks that were sold under the brand names of Bayard and Reveille-Matin. In 1922, Robert Duverdy took over control of the company and increased the sale of its products into both the home and export markets. At one time it produced a regulateur de Paris clock employing the familiar pendule de Paris movement.

The Bayard factory was still standing in 1986 (Fig 2.10), producing carriage clocks and a range of cheap battery-driven wall clocks.

Saint-Nicolas-d'Aliermont in 1889

81 Tripplin.
Watch and Clockmaking at the Paris Exhibition of 1889: Being an Account and Comparison of the Exhibits in the Horological section of the French International Exhibition. London 1889.

Julien Tripplin's[81] account indicated that by 1889, the year of the French Exhibition in Paris, the horological industry was well established in Saint-Nicolas. Of a total population of some 2,500 people, 1,500 were employed in one way or another on horological work. Many were outworkers, who combined the making of small parts at

27

home with the pursuit of an agricultural life. There can be little doubt that movements destined for marble clocks were made here, and possibly assembled into cases at the same time.

In 1889 Saint-Nicolas was producing between 250,000 and 300,000 timekeepers alone, with a total value of between £100,000 and £120,000, staggering figures for the period. From the village came chronometers, alarms, carriage clocks, turret clocks and most importantly, chimney pieces. A chimney piece is the old French equivalent of what we would now call a 'mantle clock', designed to sit on a ledge above the fireplace.

The End of an Era

The First World War had a great impact on Saint-Nicolas and by the Second World War the horological industry was a shadow of its former self. In 1986 Reveils Bayard appeared to be the sole remaining clock factory.

The town has not forgotten its past and a horological museum in the town hall provides the visitor with an insight into the wealth of material that the area produced.

The Competition

Many inferior copies of French marble clocks were made by the English, the Germans, and the Americans. At the same time some high quality pieces were produced by the English. It is possible that the original marble clocks were English and the French copied the idea and extended it into a style now considered to be of French origin. This happened at about the same time, but in reverse, with the brass skeleton clock which was French in origin, but was made in such numbers in England that it is usually considered to be an English style.

English Marble Clocks

It is normal to associate marble clocks with France and the latter half of the nineteenth century, although the English were producing superb examples as early as 1807. The makers of the high quality English marble clocks were recognised craftsmen who jealously guarded both their reputations and their customers. They are

Fig 2.11 An English marble clock of c1850 in the Egyptian style. The gilt style is signed Webster, Cornhill and the movement has a fusee. Photo: Sotheby's, London.

known to have included Arnold, Barraud, French, McCabe, Vulliamy, and Weeks.

Thwaites and Reed's embryo factory turned out many movements to be fitted by other makers into marble cases. There was great demand in nineteenth century England for French style clocks and there is little doubt that Belgian black marble cases were imported to be fitted with English movements.

The English Movement

English movements tended to follow the classic tradition of high quality, and invariably used a fusee instead of the typical French going barrel (see Chapter Four). The use of a fusee enables a quick check of the clock's origin without having to look inside the rear of the case. The winding holes of a striking movement fitted with a fusee movement are high up the dial and often line with the dial hole. This contrasts with the lower position of the French winding holes.

The English Case

English marble cases differed in construction from their French counterparts. A French case is usually made up of pieces of marble glued to a plaster base and one case could consist of many separate pieces.

In contrast, an English case was usually made of three blocks, the centre hole being drilled through a solid piece of marble. It is known that there were marble clock case makers working in Red Lion Street in London from 1800 onwards, and that a supply of black marble was available from various quarries in England.

It seems likely, however, that few cases were made from English marble and that those were only made at times when either trade restrictions were imposed, or during wartime. The Belgians produced top quality cases at very competitive prices that would have made the production of English cases unrealistic.

Some cases were made in Derbyshire during the latter half of the nineteenth century but this type of clock is fairly rare and would appear to have been sold in a very localised market.

French Clocks with English Markings

Quite often, a typical French marble clock will have an English name on the dial such as Benson or Payne & Co. These companies bought complete clocks and had their names written on the dial. Similarly many other large retail jewellers had their names recorded on the dial, buying the movements from France and fitting them into marble cases that may have been produced in Britain. Unfortunately little information is available on this subject. It is a common and understandable mistake for the novice to think that an English name on the dial

Fig 2.12 An American Ansonia clock. Note how much lower the winding holes are placed on the dial in comparison with a French marble clock.

Fig 2.13 An American Ansonia marble clock that is typical of the many thousands imported into Great Britain towards the end of the nineteenth century.

means that he has picked up an English clock, but he should be wary and look in the back first. The back plate of the movement will give an instant guide to the origins of most clocks.

The English Reproductions

In addition to producing top quality marble clocks, the English also supplied clocks fitted with poor quality movements. Some of these are reminiscent of alarm clock movements and others are similar to the construction associated with American clocks.

Around 1910, Hirst Bros of Manchester issued a catalogue showing a variety of black marble and green onyx clocks. Any case was available with one of three types of movement, namely, HAC timepiece, HAC strike, or French strike. The relative prices are interesting, being 33 shillings (£1.65), 40 shillings (£2.00), and 50 shillings and sixpence (£2.52½).

HAC are the English initials for the German movement manufacturer, Hamburg Amerikanische Uhrenfabrik. This company produced poor quality movements, using a mass-production system based on those of the American factories.

The British United Clock Company was founded in Birmingham in 1885 to counteract the massive number of clocks and movements that were imported from France. In 1889, at the French International Exhibition, they competed against a formidable array of French manufacturers with a selection of drum timepieces that were fitted with lever escapements, i.e. without pendulums. They received awards in Adelaide in 1887, and Melbourne and Sydney in 1888. Nevertheless, the business closed in 1909, some authorities attributing its demise to the stiff competition from German concerns.

American Marble Clocks

The Americans produced imitation marble clocks in their thousands, using mass production techniques that in quantity, but not quality, surpassed those of the specialists in Europe such as Japy Frères. American 'marble' clocks fall into three main categories; namely marble, cast iron and enamelled wooden cases imitating marble. American clocks are easily identified either by looking in the back door of the case, or by the low position of the winding holes on the dial (Fig 2.12). Many were imported into Britain. (Fig 2.13).

Large quantities of inexpensive American movements helped satisfy the large demand that existed both in the United States and in Europe. The State of Connecticut north-east of New York was to become the home of some of the largest clockmaking companies in the USA. Rivers were available for water power and local mineral deposits helped Ansonia and Waterbury to become famous names.

The American movements (Fig 3.11) were not particularly attractive, but did the jobs required of them. The plates were skeletonised to save metal and the springs were open, that is, not enclosed by the brass barrels associated with French movements. American springs are a potential hazard to repairers as they have to be bound with a circle of steel before the movement can be taken apart, otherwise they will 'explode' and possibly cause harm to the repairer.

Pinions are of the lantern type (Fig 2.14), which were cheaper to make than the solid pinions used in French movements. Lantern pinions may look rather flimsy, but they are not; they are successfully used in large turret clocks. The gear action between a lantern pinion and a correctly made driving wheel is excellent, although lantern pinions make poor driving wheels.

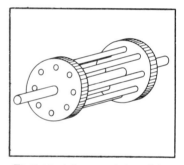

Fig 2.14 A lantern pinion of the type found in many American marble clocks.

Another feature that makes the American movement unique is the way in which the striking train is activated. Sometimes it is possible to see evidence of rack or locking plate striking, but the mechanism will often be mounted between the plates, causing a repairer additional problems.

In my opinion, they should be left alone by the novice. The complexity of the movements and the rudimentary nature of their construction pose difficulties that will often dishearten an inexperienced worker. Any novice reader determined to repair one, should read a few books that have sections dealing with alarm clocks as there are many similarities between the two types of movement. The section dealing with alarm clocks in Donald de Carle's 'Clock and Watch Repairing' was of use when I first worked on an American movement. There is also a section on alarm clocks in de Carle's later book, 'Practical Clock Repairing' (N.A.G. Press), which is still in print.

8 Day, Half-hour Strike, Gong.
STAR MOVEMENT.
Height 10⅜ inches.   Length 13 inches.
Porcelain Dial 5½ inches.   Visible Escapement.

8 Day, Half-hour Strike, Gong.
STAR MOVEMENT.
Height 9¼ inches.   Length 10¼ inches.
Dial 5¼ inches, Porcelain or Fine Gilt.
With or Without Visible Escapement.

Fig 2.15 A clock from the Waterbury Clock Company catalogue of 1891. The case was probably made in Belgium marble with a movement of American origin.

American 'Marble' Cases

The Americans imported some marble cases from Europe and fitted their own movements. The Waterbury catalogues mentioned below do not carry French marble cased clocks before 1885. The cost of transport was high and many factories produced cases that were made of cast iron lacquered to look like marble.

Iron cases have an appeal of their own to some people and in years to come many will attract more attention than they are enjoying at the moment. They are often encountered in junk shops and on market stalls. At first sight they appear to be marble, but a closer inspection reveals their metal origins. The easiest way to check for an iron case is to place the back of your hand against the case; iron feels colder to

Fig 2.16 A cast iron Waterbury clock (the 'Leona') that could easily be mistaken for a French marble clock, 1891.

Fig 2.17 A Waterbury clock (the 'Moscow') made of polished wood and fitted with an American movement.

8 Day, Half-hour Strike, Gong.
Ord. Dial, White, Black or Gilt.
STAR MOVEMENT.
Height 11¼ inches.   Length 8⅞ inches.
Dial 5 inches.

Fig 2.18 An example of a German marble clock movement. Such movements had thin plates and were usually fitted with lantern pinions.

the touch than marble. Turning the case upside down will allow you to look at the inside; also check the base for tell-tale rust marks.

It would be easy to believe that the Americans had invented the cast-iron case, but a picture of a French clock in Tardy's[80] book, 'La Pendule Française' Vol III, shows a case made of cast iron from the Empire Period of 1800 to 1830. It is in the Egyptian style and heavily decorated with gilded bronze.

Also encountered are 'marble' cases made from tin plate that have been cleverly painted to resemble onyx or a similarly marked marble. With one of these, you only have to pick it up to realise how light it is compared with marble or iron.

American Manufacturers - Ansonia

It is generally agreed that more clocks by Ansonia will be encountered in Britain than by any other American manufacturer. The Ansonia Clock Company specialised in producing either marble or imitation marble clocks in direct competition with the large numbers of French clocks that were imported during the last quarter of the nineteenth century. The Ansonia clock catalogue of 1880 shows a range which includes the familiar cast iron variety. A typical Ansonia clock is shown in Fig 2.13.

The company started life in a brassworks in Ansonia, Connecticut. In 1878 the clockmaking business broke away and moved into premises in Brooklyn, New York, that were destroyed by fire two years later. Within a year, a new factory had been built and, by 1883, all the work from the Connecticut factory had been moved to Brooklyn.

Waterbury Clocks

80 Tardy.
La Pendule Française Vol III.

20 Waterbury Clock Company Catalogue. No.131. American Clock and Watch Museum, Conneticut, USA.

The Waterbury[20] catalogues show the wide range of clocks produced, with detailed pricing of the various types. A classic French marble clock is No. 2809 shown in Fig 2.15. This clock, with all the others mentioned here, was fitted with the Waterbury Star movement, one that was considered to be better than most. It employed lantern pinions fitted between rectangular, skeletonised plates. The rack striking mechanism was fitted outside the rear plate. This clock, with a French-style marble case, cost $22.00 and in 1893 a surcharge of 50 cents was made for a visible escapement.

By comparison the Leona in Fig 2.16, fitted with the same Star movement, was made of cast iron and cost $10.50 in 1893, with a surcharge of 50 cents for a visible escapement model. The Moscow clock shown in Fig 2.17, made of polished wood and fitted with the ubiquitous Star movement, cost $7.50 in 1893. Again a visible escapement model was supplied for an extra 50 cents.

A marble cased clock of European origin would have cost twice as much as one with an American iron case, and three times more than one with a polished wooden case.

German Marble Clocks

The Germans appear to have manufactured fewer marble clocks than the Americans and with less variety. The German industry had its roots in the Black Forest, where many small workshops produced clocks of various types that were mainly exported to Britain.

Production increased until the year 1840, after which the Germans began to feel the effects of strong American competition. The industry reorganised itself into larger factory production units and to a greater extent modelled itself on the American system. Some of the larger names to emerge were Junghans and the Hamburg Amerikanische Uhrenfabrik (HAC).

German movements are usually similar in appearance to the French, with round brass plates, but the metal parts are thinner and the fittings cruder. Looked at more closely, their lantern pinions and other features are similar to the American type. They are certainly easier to work on than the American versions, but do not at this time command any higher value.

In many ways, American movements are an honest cheap alternative, whereas the German variety appear to be deluding the customer into believing that they are of French origin. In a German movement (Fig 2.18) the plates are thin and the pendulum suspension tends to be crude, but to the untrained eye they look remarkably similar to the genuine French original.

The Treaty of Frankfurt in 1870 allowed unassembled German clock movements into France free of all import duties. This is said to have created problems for the major manufacturing centres of Badeval and Saint-Nicolas d'Aliermont. The movement shown in Fig 2.18 has the words "Made In France" on the back plate although the movement is typically German with lantern pinions and poor component parts. This movement was obviously shipped in pieces to be assembled in France, fitted into a marble case there, and legitimately sold there, or exported as being of French origin.

III

How to Buy, Examine and Date a Clock

During the latter half of the nineteenth century, French marble clocks were sold in tens of thousands and enjoyed a widespread, almost universal popularity. Unfortunately their appeal waned after the first World War, rapidly becoming disliked by both the clock collecting community and the general public. The years from around 1980 onwards have seen a modest revival in their popularity and prices have begun slowly but surely to rise. There are, however, many people who still regard these clocks as a relic and reminder of the worst aspects of the Victorian era. Fortunately this attitude is changing, following the now general acceptance of Victorian art and design.

Fig 3.1 Magnificent example of the French marble clock as an art form.

It is foolish to suggest that some of the cases, such as those that take the form of black Greek temples, are in any way works of art. They can, however, add an interesting focal point to a modern room, while other cases have artistic merit (Fig 3.1). In an age of digital clocks and non-stop radio time checks, there is comfort to be gained from the solid and dependable striking of the hours by an old, homely clock that may have been steadily ticking away for the last 100 years or so.

Apart from being the cheapest of the collectable timepieces, marble clocks have a great advantage over some of their alternatives, such as carriage clocks. Their movements are reasonably accessible and a simple repair can often turn a junk shop 'no hoper' into a good working clock. The relevant chapters in this book will help owners who wish to undertake major restoration work, providing detailed information for both the experienced worker and the novice.

One reward for undertaking this often demanding work, is the satisfaction of finishing with a working clock that will be a constant reminder of the success of your endeavours. The other obvious bonus is that you will save a considerable amount of money. For a clock with a small movement, the French marble is suprisingly expensive to have repaired professionally.

Another joy of owning an old clock is that it works for its living, albeit sometimes in an erratic and temperamental manner. This is in marked contrast to 'static antiques' such as porcelain and glass that are beautiful but not always useable in day to day life. My first French marble clock was from a junk shop. At the time, it was the cheapest antique clock available that had a top quality movement. Many clock collectors have started with marble clocks for this reason alone. Usually, marble clock owners are not too proud to admit the junk shop origins of some of their treasures. It is both interesting and profitable to go into every junk shop you come across. Sometimes you may be lucky and come out with an attractive clock for a good price, or perhaps buy a damaged movement from the bottom of a tea chest full of rubbish.

Buying for Investment

There is some investment potential in the French marble clock. Stories are often heard about the apparently worthless old object found in the loft that turns out to be worth thousands of pounds. It is also true that in recent

Fig 3.2 A small, attractive French marble clock with nautical embellishments. Clock from The Clock Shop, Weybridge, UK.

years the prices of some clocks have risen dramatically. There was a time when the value of the much-cherished grandfather, or longcase clock, to give it its original name, doubled in just two years.

Newcomers may be thinking that this trend will happen to French marble clocks during the next few years, or that they will find the odd clock that is going to be resold for a handsome profit. It is wise to keep the market in perspective, bearing in mind that an article is only worth what someone else is prepared to pay for it. There are dangers in hoarding such clocks with a view to making a large profit at a later date. Some of the dangers are noted below:-

1. Remember that the clocks were made in their thousands. Those that make good investments are often those that were produced in relatively small quantities.

2. The French marble clock is continental. A British collector will often prefer an English-made clock. This does not necessarily imply a chauvinistic attitude; at one time the English made the best clocks in the world. A British collector still has access to some fine, if

41

expensive, examples of English workmanship. Another factor is that some French clocks are considered to be rather fussy for the more conservative English taste.

3. French marble clocks are considered by many people to be modern although many of them are nearly a hundred years old. Remember that clocks twice, or even three times, this age are available to collectors.

4. They do not have a universal appeal. There will probably always be people who dislike black marble with its associated images of death and mourning.

Fortunately there are some notable exceptions to the points just made.

It is likely that there will always be a market for the unusual clock, such as that illustrated in Fig 3.2, which is a nice small striking clock in a fairly ordinary case. One could be forgiven for thinking that the nautical embellishments were added as an afterthought - a retirement present for a sea captain perhaps? These features do however make the clock

Fig 3.3 Large and impressive clock fitted with the Brocot calendar movement and barometer.
Photo: Sotheby's, London.

Fig 3.4 Excellently restored marble clock showing a Brocot calendar movement. Note the lunar phases displayed at the top of the lower dial.

attractive. If the embellishments were removed, it would merely be a run of the mill movement in an ordinary case and of less interest.

Similarly the large marble clocks illustrated in Fig 3.3 and 3.4 are equally in demand. They are fitted with the relatively unusual feature of the Brocot calendar; such clocks will almost certainly remain popular and the high prices being paid at auction for them now testify to this argument.

What to Buy

When I began buying French marble clocks, I acquired nearly every clock that my meagre budget could accommodate. Having little practical knowledge, I purchased each clock for what I called my 'scrap price'. It was an entirely arbitrary figure given to a clock that was not working, and was the value I thought it would be worth if I could not get it going.

Fig 3.5 Detailed view of the popular Brocot visible escapement.

This may appear to be a strange, almost fatalistic approach, but there was some logic behind it. After all, why pay a substantial price for a clock that may never work? Fortunately, my confidence and expertise increased after I had bought about six clocks. I am now able to buy with a degree of confidence, but have never forgotten the old scrap price principle.

Types of Movement

The prospective buyer has a wide variety of cases to choose from; the movements provide little variety, but this is more than compensated for by each movement being a small work of art in its own right. The uniformity of design can be a disadvantage from the point of view of the collector, but is a great bonus for the amateur restorer who can quickly gain specialist knowledge on a limited range of movements.

The movements fall initially into two groups. The first group, timepieces, do not strike; the second group, striking movements, strike the hours and half-hours on a gong or bell. Striking movements fall into two more groups, those with rack striking, and those with locking plate striking. Striking work is considered at length in Chapter Six. From a buyer's point of view there is little difference between the two striking movements, apart from those with locking plates being slightly easier for the novice to work on.

Some clocks have peripheral functions that

Fig 3.6 The classic round - plated pendule de Paris movement that will be found in most French marble clocks.

Fig 3.7 The less common square-plated pendule de Paris movement.

Fig 3.8 Rear door view of a timepiece marble clock movement.

add interest and variety. The most popular is the Brocot visible escapement (Fig 3.5). It is a part of the movement that has been mounted on the dial instead of on the movement inside the case, hence the name. The anchor-shaped escapement is seen rocking to-and-fro with the motion of the pendulum, while the escape wheel turns in small jumps as its teeth are alternately caught and released by the ends of the anchor arms. This is an attractive feature on any movement and marks the first significant step towards the clock having a broader appeal among buyers.

The Cases

The generic term 'French marble clocks' is strictly incorrect because cases made of genuine marble were usually restricted to the early part of the nineteenth century that preceded the large volume production of the traditional black-cased clocks.

The vast majority of traditional cases are made of a material known in the marble trade as 'Belgian black marble', and sometimes called 'Belgian limestone' in horological circles. I prefer to call it 'Belgian black marble' for reasons that are discussed in Chapter Ten. Some authorities still refer to the material as 'Belgian slate', which is incorrect.

Some black marble cases will tend to look like slate when their polish has aged or has worn away, to show foggy white patches coupled with shiny black areas. On a square-sided case there will sometimes be wear on the left hand side where the owner's left hand has steadied the case while winding. On clocks with a circular surround, the patch will usually be on the top of the case.

Points to Note when Buying

The novice will need to exercise a degree of caution when buying, although there will be little to worry about if the price is very low and falls into the scrap price category.

Fig 3.9 A rack stricking movement, viewed through the rear case door.

A higher price will require careful consideration about the quality of the movement and case, together with an assessment of the honesty and credibility of the vendor.

Assuming that the clock is being bought from an antique dealer or junk shop, the first point to check is that you are actually looking at a French marble clock. This implies that the movement must be a top quality version of the pendule de Paris calibre, Fig 3.6. Occasionally you will come across clocks fitted with the equally good square-plated movement, Fig 3.7. The only means you have of checking this is by opening the little back door on the case and peering inside. Fig 3.8. and Fig 3.9 show the kind of view that you will have. Ideally, of course, the movement should be removed from the case for inspection, but in a shop this will usually be impractical.

Fig 3.10 An example of a typical movement maker's stamp that often appears on the back plate of a movement. It is about 10mm in diameter and an indication of a good quality movement.

If a bell is mounted on the back of the movement, it will have to be removed by unscrewing the little knurled knob that holds it on the bell post. Most shop owners

46

will be happy for you to do this. If they are not, it would be wise to explain why you wish to remove it. Underneath the bell there may a little manufacturer's stamp; an example of one of these is shown in Fig 3.10. This is a good sign and will indicate that you are probably looking at a movement made by one of the better makers of blancs roulants.

Many such companies won awards, not only for the quality of their movements but for the quantity produced. In 1855, the Japy company produced 60,000 blank movements; this large volume however does not necessarily indicate a diminished quality of production. The French horological industry worked to exacting standards set by the Parisian finishers of the movements. The little stamp therefore provides a good guide to begin with.

Regrettably not all movements were stamped, and many a fine one is found to be unstamped. Several of my clocks with perfectly good movements have merely a few numbers stamped on the back plate. The implications of these stamps and other indicators are examined later in this chapter.

Beware of Imitations

Fig 3.11 A movement from an American marble clock. Compare its utilitarian construction with the elegance of the French movement in fig 3.0.

The movement shown in Fig 3.11 is from an American marble clock. Taking advantage of the enormous popularity of the French marble clock, the American and German industries made inexpensive copies in large quantities during the latter half of the nineteenth century. These were considerably cheaper than the French models and were sold in their thousands. They have their own peculiar characteristics, which were were noted in Chapter Two. From a practical point of view, the movements are of an inferior quality and can be particularly awkward to work on.

Fig 3.12 shows a movement with a stamp on it, and the words 'Made in France' on the back plate. (It is just visible in the illustration between the horizontally projecting straps). It should, however, be treated with caution; the clock was made during the beginning of the twentieth century. A Britannia movement is indicated, implying that it was made specifically for the British market. The pendulum suspension at the top of the back plate is of an inferior type. A look inside the movement confirms the poor quality more associated with American clocks.

Fig 3.12 Although stamped 'Made in France', this inferior movement was made in Germany and assembled in France.

To conclude the initial inspection, it is therefore necessary to check the following features on the back plate:-

1. A small stamp or medallion on the back plate bearing a name and such words as Médaille de Bronze, will usually be a good guide to quality.

2. The plates should be held together by steel pins and not nuts.

3. Stamps on the back with English sounding titles such as 'The Brittania Movement' or 'Made in France' should be treated with caution.

4. Check that a pendulum is fitted. Most French marble clock movements used pendulums, not platform escapements, which have balance wheels and hairsprings like watches.

5. Plates should be of brass. Beware of any that appear to have been lacquered or plated; some imitation movements have thin, electroplated, plates.

6. A suspension of the type shown in Fig 3.12 is of inferior quality. The presence of a pendulum suspension of the Brocot or Vallet type will be a useful guide to the overall quality of the movement. (See Fig 5.16 and 5.17).

As a final word about movements, in the long run, experience is the only guide. The newcomer should play safe by trying to buy a clock at a low price or one with a movement that has a stamp on it as illustrated in Fig 3.10. An alternative is to find a reputable dealer who sells only fully-restored clocks.

What Type of Case?

The appeal of the case will be based upon personal taste and preference. To some readers, the mechanical aspects will be secondary to the overall shape and appearance of the case. Do you like the clock? Try to judge the size of a case in relationship to the size of your home. In a large shop, the cases tend to look smaller than they really are, so be careful not to buy a large piece that may dominate the room of a small house.

If you are on good terms with a local dealer, he may allow you to take the clock home and try it for size. The obvious alternative is to measure the case and find an equivalent sized box at home to try where you intend to stand the clock. This may sound trivial, but one of the most frequent criticisms levelled against marble clocks is that they are 'big and ugly'.

The overall condition of a case is an important factor and is often overlooked in the confusion that often prevails while looking at a clock. I am not particularly worried if a case has an inconspicuous chip out of it, but some friends regard any blemish as an unsightly. This will be of little concern if you intend to keep the clock, but may pose difficulties if you are intending to sell it at later date.

While assessing condition, be on the look-out for hairline cracks that indicate the case has been dropped or badly damaged. This often happens because the clocks are very heavy and many owners have a dangerous habit of picking up the case from the top, instead of lifting and carrying it from underneath.

Watermarks are a common nuisance on a case, often taking the form of white rings on the flat tops of cases made by wet drinking glasses. They can be particularly difficult to remove.

49

Case Inspection Summary

1. Check to see that the case has not been dropped.

2. See that the extremities are not too badly chipped.

3. Make sure that a large case is not too big. If you take a liking to a clock that is too big for a living room, remember that it could be placed on a bracket on the hall or landing wall.

4. Look carefully for watermarks on the case, especially where wet glasses may have been placed on the top.

5. Examine the dial, bezel, and hands.

Dials will sometimes be chipped around the winding hole. Personally I do not mind a slightly chipped dial, but dislike one that has been badly retouched with a paint that has yellowed with age. A cracked dial may, of course, be a good reason for giving the clock a wide berth unless the price is low, or you are confident that an old replacement dial may be located. The cost of repairing dials is high and sometimes there is little that can be done for one that is severely damaged.

The winding hole inserts may be missing. Some inserts are made from blued steel; others are of brass.

The bezels are the brass surrounds to the dial and the front and rear door glasses. The main problems encountered here will be cracked glasses and brasswork overpolished by a previous owner; sometimes this is in keeping with the original design of the clock, but too often the brass has been polished unnecessarily.

Flat glasses are available very cheaply from clock material dealers, but the thick bevelled types will have to be made to order and will consequently be more expensive. Little can be done about a case with gleaming brasswork, except to take note of the comments on the subject in Chapter Twelve.

Broken hinge pins on doors do not pose any great problems, although some surgery may be required if the brass work of the hinges themselves is damaged.

While looking at bezels, peer inside the rear door of the case to see if the straps (Fig 3.13) appear to be original. There is a good chance that the movement itself

Fig 3.13 Detailed view of a rear case bezel showing one of the movement straps and its retaining screw.

is not original if the straps are particularly short or have been tampered with. This will not normally pose problems unless the replacement is totally out of keeping with the case, or when the pendulum requires a hole in the base of greater width or depth than is provided in the replacement case. Movements are often swapped around by shop owners to provide working and saleable clocks. Do not worry too much about such 'marriages' with cheap clocks; use the fault to your advantage as a means of reducing the price.

The main points to consider with hands is whether they are original and if they have been damaged and badly repaired. Steel hands should ideally be a fine oily blue colour, but are easy to reblue should the need arise.

The clock should be supplied with a key, preferably the original. These are not particularly expensive items, but are often forgotten at the time of purchase. On some movements, a small extra key will be required for the little square arbor at the top of the dial; this adjusts the rate of the movement. Sometimes you will be lucky and come across an original key.

Main Items to Keep in Mind

1. Dials should be free of cracks and serious chips.

2. Ideally bezels should have sound glasses and hinges.

3. Check to see that the hands are original(see pp.77-78) and that they have not been badly repaired.

4. Are there winding keys?

Guarantees

I have been buying clocks for several years and have been assured by some vendors that there has been a guarantee on everything, from a splendidly restored clock to a tortured heap of clock mechanism supposed to be a 'good working movement'.

The main point to bear in mind is that genuine guarantees cost the vendor money. For example, at the time of writing the cost of fully restoring a French movement is more than the price I am currently paying for a small marble clock of good quality. Working on a proportional basis, the cost of a simple repair would probably be in the order of a quarter to a fifth of the purchase price.

If a dealer is offering a genuine one year's guarantee on a clock, he will have to insure himself against the risk of that clock coming back for repair during the guarantee period. This insurance will usually take the form of a higher initial retail price to the customer. If it does not, something is wrong. Many specialist clock dealers will not sell any piece that has not been carefully restored, preferably in their own workshops. A good dealer's time and reputation is too valuable to be wasted on unreliable movements.

Some people will, without making any attempts at restoration, offer a clock for sale in the hope that it does not come back during the guarantee period. I have known some shopkeepers who operate on this basis; if the clock comes back, they may offer the owner a refund rather than bear the cost of a repair. The worst situation arises when the clock goes back for repair and sits for months gathering dust on a shelf in the back of the shop.

The moral is that 'you only get what you pay for'. If you trade cheaply with junk shops, do not expect the same after-sales service that is obtained from a bona fide dealer who charges more for the same clock. My personal approach is to ask the seller if a guarantee is given; if the answer is "Yes", I ask for the guarantee to be waived and a discount given on the purchase price. Many shop owners will be happy to do this.

If you do opt for a guarantee, make sure that your written receipt includes the words, 'One year's parts and labour guarantee given on this clock'. It may sound cynical, but I have little faith in 'guarantees'. They are seldom worth considering as a positive attribute of a purchase, with the exception of those given by reputable specialist clock dealers.

Summary

1. Do not expect to buy cheaply and obtain watertight guarantees.

2. If a guarantee is given, get it in writing, preferably on the receipt.

3. Try to have the price reduced in return for a 'sold as seen' sale.

4. Do not expect a junk shop to give the same after-sales service that an accredited specialist will provide.

Where to Buy a Clock

There are three principle sources of clocks: shops, auction houses and market stalls. There are also three types of shop: specialist clock dealer, antique shop, and the ubiquitous junk shop, where the most surprising and interesting finds may be made.

Junk Shops

'Doing the junk shops' can be hard work, to have some degree of success, it may be necessary to visit every one you come across. It is worth the effort because bargains are often discovered. There inevitably will be shops where a simple marble clock has been grossly overpriced, but do not lose heart. Make a sensible offer for the clock. If it is refused, keep going back once a month. The chances are that it will not have been sold and the shopkeeper may eventually accept your offer. Few shop owners will mind if you leave your telephone number and ask to be given a call if they change their minds. Remember that these comments will only apply to junk shops. Antique shops may be open to offers, but many will express shock that you consider their stock to be of such little value.

The Specialist Clock Dealer

A different situation exists when a dealer specialises solely in clocks; apart from having a good working knowledge of the trade, he will also have a good idea of the market value of his stock. In many shops the clocks have been fully restored and are sold with a genuine guarantee. This high standard of service will be reflected in the prices asked. If you wish to make an offer for a clock, do so by all means, but do not appear to be working the price down for the sake of it. The best approach is to be honest. You may feel that some extra work needs to be done or you might not have wanted to spend so much money. Above all, do not be in too much of a hurry to buy. You can always go back later after giving the item some careful thought.

Market Stalls

There are many bric-a-brac markets scattered all over the country. On the continent, places such as Paris and Vienna have always had their famous flea markets. Germany has some big markets in Berlin and Hanover displaying fine clocks, although in these you will be more likely to find German imitations of French marble clocks.

Bermondsey Market in London is an extensive antiques market that opens very early on Friday mornings for trade buyers and changes to retail dealing as normal business hours approach. Covent Garden market in London no longer sells fruit and vegetables but has a bric-a-brac market on Mondays.

Buying a clock from a market stall is an experience in itself. Some stalls are run by dealers with long experience of their respective trades; one of England's current clock specialists once ran a stall at Bermondsey. Other stalls are run by opportunists who buy and sell. These people are usually looking for a quick sale; I have often stood looking at a fine marble clock at Covent Garden while the price continues to be lowered in large steps.

It is suprising how often a polite and quiet approach towards the stall holders will help with bargaining for a good price. I usually ask the price immediately and then spend a long time looking at the clock, peering in the back and removing a bell if necessary. Sometimes the stallholder will lower the price without any further prompting.

If you wish to make an offer, the following way is often the best. Suppose you have seen a clock that costs £100. Ask the stall holder if he 'can help on the price' or 'is open to offers'. He will almost invariably say 'Yes'. At this point, make an offer that is lower than you would want to pay. For the £100 clock, offer say £75. The seller will probably now try to tell you that the clock cost him more than that. You can now save his pride by asking if he 'will meet you half way.' This approach is often acceptable and you will buy the clock for about £90. Try to remember that the stallholder is human and that you must consider his or her self-respect and dignity.

Auction Houses

Many auctions take place around the country, varying in size from small country events held in church halls to the larger auction houses in London bearing internationally known names. Some auctioneers have sales at regular intervals that are specifically for old clocks and watches. Catalogues are usually available for a fee ranging from a few pence to several pounds. The best way to keep track of these sales is to look at one of the quality newspapers, such as The Times, where the notices usually appear on the court and social page. Alternatively, it is possible to subscribe to a mailing list that will give advance warning of future sales. For provincial sales, the local paper is the best place to look for notices.

To buy successfully at auction, you will need to know the current market prices and have a good knowledge of your subject. For the newcomer, viewing days provide a marvellous opportunity to see all sorts of clocks 'in the flesh'. If you are fortunate enough to work near some of the biggest London salerooms, although you may rarely buy there, it is rewarding to make an effort to go to viewing days. At the time of writing, the only marble clocks that were making their way into the bigger London auctions were the of more splendid type illustrated in Fig 3.3 and 3.4.

The joy of going to a viewing is not just to hunt for marble clocks, but also to look at all sorts of timepieces, from longcase clocks to fine old watches. On entering one of the larger salerooms, you will often find several rows of tables with seated men prising open the backs of pocket watches as if they were oyster shells; they are dealers checking the movements before making an assessment of the price to pay on the day of the sale.

Most of the people encountered will be from the antiques trade, but do not be intimidated by this.

Auctioneers usually like to see the private collector competing with the dealers. Do not hesitate to examine the clocks that interest you. If necessary, ask a member of the staff for advice; they are usually only too pleased to help. Some auction houses take the movements out of the longcase clocks and place them in front of their respective cases. This provides an ideal opportunity to note the different techniques employed in the construction of these ever popular clocks.

Buying at Auction

The fundamental idea behind an auction is that an item is sold to the highest bidder. The conditions that apply to the sale are governed by the law of contract. In theory, it is possible to claim compensation against an auctioneer who makes a false statement on a product. For example if the catalogue states that a case is 'made entirely of solid marble', this would be a false statement if the case were later found to be a cast-iron imitation.

The auctioneer should say something like, 'This case is believed to be of solid marble'. You will often see a similar statement along the lines of, 'a fine timepiece attributed to Pierre Le Roy'. This is not the place to go into the legal complexities of buying by auction; the overriding factor is that you should make up your own mind whether or not you can afford a clock and whether you believe it to be genuine. The old principle of only buying something that you like still applies; if you make a poor buy, you should at least be able to live with it.

At auction, the prices paid will normally be lower than those for clocks found in shops. The sales are in effect at trade prices. The vendor of an item pays the auctioneers a commission, and sometimes the buyer has to also pay a premium.

Making a Bid

Many hair-raising stories are told about the innocent spectator at an auction who scratched his head, only to find that by doing so he had inadvertently made a bid of thousands of pounds for something he had no intention of buying. Most people bidding will raise a hand or gesticulate with the sale catalogue. If you do not intend to buy anything, it is safest to keep away from the centre of proceedings.

It is usually possible to leave a bid on an item whether or not you attend the sale. A tip to bear in mind is not to leave an offer with a rounded number such as £300. It is better to offer £310. Many other offers of £300 may have been received. Yours, being slightly higher, will be accepted.

Reserve Prices

One of the frustrating aspects of a sale by auction is that a minimum price may have been set on the item for sale. For example, a clock may have a reserve on it of £300. If this price is not met, the item will be withdrawn. Unfortunately reserve prices are not published, although there is an intention to change this state of affairs.

Buying from Auction - A Summary

1. Do not buy unless you are fairly sure of your subject.

2. There are bargains at auctions although they are the exception rather than the rule.

3. Remember that you can make a postal bid if you wish, or leave details of your offer at the auction house before the sale starts.

4. Remember that clocks sold at auction are usually 'sold as seen'. You will have little recourse if anything is wrong with the movement. The price you pay, however, will usually be considerably lower than the normal retail price.

5. Seek advice from the auctioneers.

Dating a French Marble Clock

The French marble clock was mainly produced during the seventy years from around 1845 until the First World War in 1914; changes in fashion and the impact of war on the French horological industry led to its demise. The First World War also marked a change in furnishing styles in England. Gone were both the sombre Victorian tones of the previous century and the appeal of black marble clock cases. Mechanical clocks with cases of light coloured woods such as oak and the later introduction of electric clocks with Bakelite cases marked the end of one of the most popular types to have been produced.

A case can provide an approximate idea of a clock's age although there always exceptions to the general guide lines provided here. The earliest cases were of very plain marble without any decoration. Intricately worked marble and volutes were introduced between about 1870 and 1880. Later models, including many American clocks, had flatter and less decorative sections of marble. The temple shape dates from around 1890 and was in evidence until the beginning of the First World War.

The main part of the clock that helps with dating is the movement. Unfortunately movements were mass-produced in large numbers to uniform designs and fitted to cases by a host of 'clockmakers', with the result that most of them have no helpful marks. Movements that are made manually by one craftsmen are more likely to have been signed than those produced in their thousands. Manufacturers like Japy produced 'blanc roulants', that is blank movements and the finishing of the product was often left to a Parisian finisher who would then ship them on to a case maker.

Fig 3.14 Back plate of a French marble clock movement showing various stamps and identification numbers. (The hammer head has been removed for clarity).

Although the following guidelines to dating are based on a large amount of research into the subject, they often express my own views and opinions. Many reference sources cannot be verified, and it would therefore be foolish to treat this as a definitive guide. I have added material that will help date any movement of this calibre back to the nineteenth century. They were so widely used that you may happen upon a similar movement in a variety of cases.

The first step towards dating the movement is either

58

to take it out of its case, or peer at it through the rear case door. The second method is not to be recommended for serious work but, when buying at auction or from a shop, this will often be the only option available.

Any information of substance will be found on the back plate. That shown in Fig 3.14 is a classic example providing a lot of information, starting with the medallion at the bottom of the plate bearing the lettering: JAPY FRERES & Cie. Gde. MED.D'HONNEUR. This of course identifies the movement as being built by Japy, one of the largest and most significant producers.

Books to Help with Dating

From here the reader could look up the Japy trademark and name in Appendix 1 of this book or turn to one of three other books. One is the well known book on this subject, 'Dictionnaire des Horlogers Français' by Tardy[5]. This is a classic work that has never been translated into English. Much of the text consists of town names and dates that are highlighted in bold type, allowing easy references for the reader who does not understand French. The only drawback is that the book is fairly expensive and only available in paperback. It is hard to find in provincial reference libraries. I know of two London libraries holding it: The Westminster Central Reference Library near Trafalgar Square and the Guildhall Library in the City, which contains the library of the Worshipful Company of Clockmakers. Local librarians have looked politely blank when I have mentioned the book in the past.

Alternatively the two volumes of 'Watchmakers and Clockmakers of the World'[21], give some information about French makers although they do not have the depth of the 'Dictionnaire' by Tardy. Volume Two was compiled by Brian Loomes and deals more with the nineteenth century than Volume One which was produced by the late G. H. Baillie. Both, published by N.A.G. Press, have the advantage of being inexpensive and would form an affordable basis to any new horological library.

Finally Charles Allix produced a book called 'Carriage Clocks'[1]; many makers of marble clock movements also produced carriage clock movements. In the back of this book is a useful list of makers' names and trademarks. It is more readable than Tardy and has the advantage that it covers other countries such as America.

5 Tardy.
Dictionnaire des Horlogers Français. Paris 1971. The standard reference work for anyone wishing to date French clocks. Unfortunately it is very expensive although only available in paperbinding.

21 Baillie.
Watch and Clockmakers of the World Vol.1 NAG Press, Colchester, England. (See Loomes for Vol.II).

1 Allix and Bonnert.
Carriage Clocks - Their history and development. Antique Collectors Club, Suffolk, England. 1981. Many makers of carriage clocks also produced the pendule de Paris movements that were used in the French marble clock. This book gives a well-illustrated and informative history of many makers, such as Japy and the companies from Saint-Nicolas d'Ailermont.

59

Dating the Back Plate

The medal in the middle of the Japy stamp (Fig 3.14) is similar to the Legion d'Honneur medal. The list of Japy's medals in the appendix suggests that the movement was made after 1888, although there are some notable exceptions to this.

The next clue is the symbol near the top on the left of the back plate in the same illustration, which carries the letters RC. This is the mark of the company which finished the movement, C. A. Richard et Cie. of 32 Rue de Bondy in Paris. The company was believed to have been founded in Paris in 1848 under the name of Lemaitre & Bergman.

From these two stamps it is possible to conclude that the blanc roulant was made by Japy after 1880, sold to Richard who in turn probably put the movement into a case before shipping it to England, where it was sold through their London branch that opened in 1857. It is interesting to note also on this plate the stamp 'Fabrique de Paris', a classic marketing ploy that would be outlawed today. The movement would, without doubt, have been made in Beaucourt which is miles away from Paris. Unfortunately, dating is not always this simple and it will be necessary to rely on more diverse sources of information.

Fig 3.15 A silk suspension; a white card has been placed behind the thread to aid viewing.

Mainspring and Pendulum

Occasionally a mainspring is seen that has been dated on its outer end; this is not uncommon on some earlier clocks. The biggest nuisance is, of course, that it is only

Fig 3.16 The Brocot adjustable pendulum suspension.

Fig 3.17 The Vallet suspension. The moving parts are more enclosed than those of the Brocot type in fig 3.16.

found during restoration. Of all dating techniques this is probably one of the safest, although one should not forget that previous repairers may have replaced a defective spring with one from another movement.

On first sight the pendulum would seem an unlikely place to look for dating information. It can, however, be suprisingly useful. Pendulums were commonly used in French marble clocks are illustrated in Fig 5.1.

Less frequently found are pendulums with a silk suspension. This takes the form of a pendulum attached to a rod with a hook on the end (Fig 3.15) and was in use until approximately 1850. It was superseded by the popular Brocot suspension (Fig 3.16), patented by Brocot in 1841, which became widely used around 1850. In 1865 Brocot took out a patent for the little spring device that gives an audible click which can also be felt as the suspension is adjusted.

The suspension attributed to Vallet (Fig 3.17) appeared around the middle of the nineteenth century. It is simpler and more efficient than Brocot's although it employs the same principle of altering the length of the suspension. One other suspension I have seen was on a Japy movement dated 1880. It was similar to the Vallet pattern but had the name 'Duvoye' stamped on the top wheel. Duvoye obtained a patent in 1852 for a type of suspension. The Thieble type appeared around 1870.

Other Dating Guides

The visible escapement often found in French marble clocks can give some idea of its date. Fig 3.5 shows just one of various types that will be encountered. Chapter Four gives more details.

With many antique clocks, the patterns of the original hands are a good indication of date. Sadly, this does not apply to French marble clocks. Only five main types were used throughout their history. Chapter Five gives more information on this subject. Striking mechanisms pose a problem to the would-be historian. The locking plate mechanism certainly predated rack striking, but the locking plate was extensively used to provide a slightly less expensive movement, even after 1900. My own inclination is to discount the striking work as a guide to dating.

The last item left to help is the signing of movements by restorers who have repaired the clocks over the years. Often they signed and dated a clock in a hidden place as a means of dating the guarantee that was given for repair. The marks will usually appear on the pendulum bob, the dial plate, or the front plate of the movement. The only point to watch is that a date shown by 12/80 could imply December 1980 as well as 1880. American readers should note the European tradition of dating by day/month/year as opposed to their month/day/year method. Spidery copper-plate handwriting may be a clue to the period, but this is not always a safe assumption. Handwriting as evidence therefore should be treated with suspicion, but not totally discounted.

IV

The Movements of French Marble Clocks

This chapter takes a close look at the various movements found in French marble clocks; a detailed study is then made of their components giving, where possible, a little historical background. The text is designed to give the reader a clear understanding of how the movements function and provide a reference source for restorers who are having difficulty with a particular movement.

The Timepiece Movement

The timepiece is the simplest of the French marble clock movements; its simplicity lies in the lack of a bell or gong to sound the hours, thus removing the need for a complex striking mechanism. The parts in a timepiece are identical to those found in the timekeeping or going mechanism of a similar striking movement.

The basic movement stripped of its front bezel and dial consists mainly of toothed wheels held between two brass plates with a pendulum hanging on the back of the movement. The theory behind the mechanism is fairly straightforward. A summary of the action will now be made, followed by a detailed description of each component.

The power is derived from a spring formed from a coiled strip of tempered steel wound inside a barrel, Fig 4.1. The spring is tensioned by being wound with a key. When wound up, it is prevented from running down again by a ratchet, called a 'click' by clockmakers. This means that the spring stays in place and turns the barrel. The barrel has teeth cut in one edge which drive a centre wheel, so called because it is in the centre of the clock. The centre wheel turns once an hour and carries the minute hand of

Fig 4.1 The mainspring, after removal from barrel (on the left). The barrel is rotated by the spring when the clock is going and has teeth to drive the other wheels of the clock. (Below) the barrel cover, which snaps into place.

63

the clock. The hour hand is driven by simple gearing from the centre wheel called the 'motion work'. There is an intermediate wheel between the barrel and the centre wheel which enables the clock to run for a week at a winding instead of a day. This wheel also enables some clocks to run for fourteen days at a winding. In simple 30-hour clocks that need to be wound daily, the intermediate wheel is omitted.

The centre wheel also drives the escapement or oscillator that is the 'heart' of the clock; an essential item to make the clock keep time. The escapement permits the spring to run down in tiny jumps, which cause the tick and tock. The rate at which the little jump occurs is controlled by a pendulum swinging to and fro, or by some other oscillator such as a balance and hairspring. The escapement also gives the pendulum a little push at regular intervals to keep it swinging. It is surprising how many people do not realise that the pendulum needs power and think that somehow it swings freely of its own accord.

Fig 4.2 Rear view of a timepiece movement with the pendulum removed and the brass back plate replaced with one made of clear Perspex. (a) mainspring barrel containing the coiled mainspring, (b) intermediate wheel, (c) centre wheel, (d) third wheel, (e) escape wheel, (f) anchor, (g) crutch, (h) pendulum suspension.

The series of toothed wheels driving the escapement and hands is the timekeeping train, more commonly called 'the going train' by clockmakers'.

In order to relate this explanation to an actual mechanism, see Fig 4.2, in which the back plate has been replaced by a sheet of transparent Perspex so that the wheels can be seen. The spring barrel, which incorporates the main wheel (driving wheel), is at *a*, the intermediate wheel at *b*, and centre wheel at *c*. The arbor of the centre wheel projects through the front plate and carries the motion work driving the hands, described later. Geared to

the centre wheel is another wheel *d*, called the 'third wheel', which drives the escape wheel *e* through the escape wheel pinion on the near end of the escape wheel arbor. Above this is the anchor, *f*. Although the connection cannot be seen in the picture, the anchor arbor is connected to the arm *g*, the crutch, which has a forked lower end at right angles to the upright.

Together, the escape wheel and anchor form the escapement. The pendulum hangs from the suspension *h* and the pendulum rod passes through the fork of the crutch *g*. As the escape wheel turns, its teeth alternately engage and disengage the anchor, rocking it to and fro at a rate depending on the swing of the pendulum, to which the anchor is linked by the crutch. The rate at which the pendulum and escapement works has been carefully calculated, along with the sizes of the other wheels in the movement, to rotate the centre wheel once an hour.

The Clock Plates

The wheels of the movement are pivoted between two plates which, for most French marble clocks, are made of a fine yellow brass that can be polished to give a deep sheen. They are usually circular, although occasionally

Fig 4.3 Front plate of a timepiece movement. (a) motion work consisting of three wheels, (b) click spring, (c) click, (d) click wheel, (e) winding arbor.

square plates are encountered. The movements are beautifully proportioned and are a result of a long and noble tradition of French clockmaking that at one time was directly sponsored by the Kings of France. The information stamped on the plates provides reasonably accurate dating of the movement as explained in Chapter Three.

Brass is an alloy of copper and zinc and the type used in the construction of the French plates consists of 70% copper and 30% zinc to give a top quality yellow brass. The wheels are usually made of a harder brass consisting of 65% copper, 34% zinc, and 1% lead or tin. The addition of lead or tin improves its handling with tools. An excess of lead however makes the surface difficult to polish and leaves a pitted and speckled appearance. It has been suggested that the thickness of the plates was about 1.7 mm up until 1810 and then increased to around 2.1 mm after that date.

Fig 4.4 The motion work, from the left: Cannon pinion, minute wheel, and hour wheel.

The Front Plate

In a timepiece, the front plate (Fig 4.3), is slightly more complex in appearance than the back plate. It carries the motion work, outlined in white and marked *a* in the picture. The motion work is the simple gearing arrangement that drives the hour hand around the dial twelve times slower than the minute hand. Fig 4.4 shows the three components: the cannon pinion (left), the

Fig 4.5 Back plate of a timepiece
movement. (a) pendulum
suspension block, (b) pendulum,
(c) tapered steel pin, (d) end of
brass pillar, (e) numbers
identifying pendulum length,
(f) movement serial number,
(g & h) where the movement
maker's mark is usually found,
(i) usual position for a movement
finisher's mark.

minute wheel (centre), and the hour wheel (right). As can be seen in Fig 4.3, the hour wheel fits over the cannon pinion, which is itself a push fit on the protruding arbor of the centre wheel of the clock. The minute hand fits on the squared end of the cannon pinion, which rotates once an hour. The cannon pinion drives the minute wheel, which in turn drives the hour wheel, on which is mounted the hour hand. The motion work provides a 12-to-1 step-down gearing. Also shown in Fig 4.3 is the clickwork that holds the mainspring in tension. This consists of a spring b, click c, and click wheel d; it is a simple ratchet arrangement that stops the mainspring from unwinding. The winding key sits on the squared winding arbor e and is turned clockwise.

The Back Plate

The back plate (Fig 4.5) is visible when you look at the back of a clock. Mounted at the top of it is the pendulum suspension block a, from which hangs the pendulum b. It is attached by three tapered steel pins, one of which is shown at c, pushed into holes across the ends of the three brass pillars d, which are riveted at their other ends to the front plate.

At the bottom of the back plate are some numbers e, which relate to the length of the pendulum and are explained in Chapter Five. The serial number f of the movement is usually stamped on the middle left side of the plate. This number will also be found on the pendulum bob and the front bezel surround.

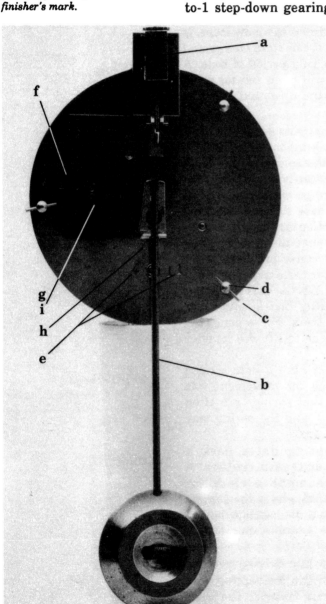

The makers' trademark or stamp will sometimes be found in one of the positions shown by *g* and *h*. Another mark, marked *i*, is that of the finisher of the movement. Chapter Three gives some more information on these marks.

Mainspring and Barrel

The mainspring in its barrel (Fig 4.1) is sometimes described as a type of motor that drives the movement. It is a rechargeable power source that can store energy and slowly release it over a relatively long period of time. A French marble clock mainspring will often run for up to fourteen days before it has become unwound and has dissipated all of its stored energy.

A separate train of gears driven by its own barrel and spring is incorporated in a striking movement. The strike spring is often similar to the spring in the going train, although this does not necessarily imply that they are interchangeable. The two springs may appear to be similar in appearance but often have different physical characteristics to suit the particular job that they perform. It is all too easy to exchange the springs when rebuilding a movement; this can cause problems of an intermittent nature that may not become apparent until the movement has been rebuilt. A bell striking mechanism will usually use a spring that is equivalent in strength to that of the going train. For a gong, the strike spring is generally stronger to lift the heavier hammer.

The mainspring will sometimes have a date scribed on its outer end to indicate when the movement was assembled. Be warned however, that springs are often changed around during restoration and the spring may not be the original one.

The history of the mainspring dates back a suprisingly long way into the sixteenth century. A clockmaker called Jacob Zech made a clock for Sigisimund 1, King of Poland, in 1525, and is the earliest known spring driven clock to have a date stamped on it. It is now the property of the London Society of Antiquaries.

French clockmakers employed this driving system, incorporating a going barrel with the mainspring in a brass barrel. English clockmakers favoured the more accurate device called a 'fusee', although it was more expensive to make. The Americans used an open spring that was not contained within a barrel, a system with a number of drawbacks that was cheap to manufacture.

Theoretical Aspects of Springs

Three factors affect the power of the spring, namely: the angle of winding, the elasticity of the steel, and the size of the spring. The bending moment of a spring may be expressed by the formula:-

$$Ewt^3A / 12L,$$

where
E   is the modulus of elasticity of steel
w   is the width of the spring
t   is the thickness of the spring
L   is the length of the spring
12 is a constant applied to a material of rectangular section.

From the formula it may be seen that the thickness of a spring is the most important factor; doubling the thickness increases its strength eight times. This high figure results from the inside of the spring being under compression when wound, and the outside being under tension, i.e. stretched. The length of a spring affects its force inversely. If a spring is shortened, its resistance to bending increases.

Fig 4.6 The anchor escapement. Only half of the escape wheel W is shown. The action is explained in the text.

When a beginner is experiencing problems with a clock, there will always be the temptation to fit a thicker and therefore stronger mainspring. Unfortunately this usually creates more problems than it solves. Always replace a spring with an exact replica, unless of course you believe that the original has been incorrectly fitted at some time in the life of the movement.

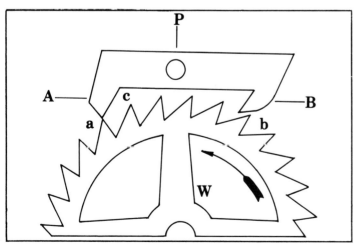

69

The Anchor Escapement

The regular little push given by the escapement through the crutch to the pendulum is known as the 'impulse'. Its function is to maintain the motion of the pendulum. In turn, the regular swings of the pendulum permit limited releases of energy by the escapement. The subject of escapements can be long and complex. This book covers only the main types that will be found in French marble clocks, with short notes about their historical backgrounds and the theory behind their operation.

The anchor escapement (Fig 4.6) appears to have been invented not long before the year 1670 by the Englishman William Clement (1638-1704). Most writers credit Clement with the invention although some attribute it to Robert Hooke (1635-1703) or Samuel Knibb (who died before 1674). The French used it after 1695 when it slowly began to replace the verge or crown wheel escapement (Fig 4.7) that had been in existence for a century. The anchor escapement made a significant contribution to the technical development of clocks by enabling pendulums to swing in a much smaller arc than with the original verge escapement. Clocks with long seconds beating pendulums - which swing from one side to the other in a second as in longcase or grandfather clocks - gave the best rates, some with variations of less than ten seconds a day.

The operation of the escapement is simple in principle. The escape wheel W (Fig 4.6) is driven in an anti-clockwise direction by the mainspring and the going train. The pendulum (not shown in the diagram) attached to the anchor is swinging to the left, which rocks the anchor about its pivot P and lifts the pallet A just clear of the tip of tooth a. This allows the escape wheel to turn until tooth b is stopped by pallet B moving downwards into its path. The pendulum is brought to a stop and the pressure of the tip of tooth b on the bottom face of pallet B rocks the anchor in the opposite direction, swinging the pendulum to the right. The action is now repeated on the other side of the anchor. Tooth c is intercepted by pallet A, stopping the pendulum from

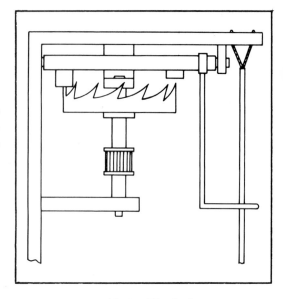

Fig 4.7 *The obsolete verge escapement.*

70

swinging further to the right and forcing it, by pressing on the inner face of pallet *A*, to swing left again. The actions are repeated as long as the escape wheel continues to be driven.

Of course, the anchor and the pendulum do not stop dead when an escape wheel tooth drops on to a pallet. The weight of the bob makes the pendulum continue a little way further in its swing before the swing is reversed. The descending pallet continues a little further in the same direction before being forced upwards. As a result, the escape wheel is pushed a short distance backwards until it overcomes the inertia of the pendulum and jumps forwards. This recoiling action may be seen by watching the seconds hand on a large movement like that of a grandfather clock. It also gives the anchor its alternative name of 'recoil escapement'.

Early anchors took the form shown in Fig 4.6 with its span of between 3 and $5^1/2$ teeth of the escape wheel. Unfortunately it could only be used with light pendulums with bobs weighing between 15 and 30 grammes. It can give good results when used with a silk suspension - a loop of silk thread on which the pendumum is hooked.

The Gable and English Anchors

The escapement shown in Fig 4.8 has the name 'gable' because it looks like the gable end of a domestic roof. It spans a greater number of teeth than the small recoil and allows the use of heavier pendulum bobs ranging from 120 to 240 grammes in weight. It has been suggested that the design was intended to allow the set-up of the anchor to be easily adjusted. One of its drawbacks is that the pallets tend to wear quickly, and in direct relation to the length of the arms.

The gable anchor was eventually modified to produce the widely-used English anchor in Fig 4.9. The French have always wondered why one of their most widely used escapements should be called 'English', especially when the English did not use it extensively in this particular form. The escapement most frequently encountered in French marble

Fig 4.8 The French gable escapement.

clocks is essentially the gable type scaled down to span fewer teeth (7½) and have shorter arms. It first appeared in such movements around 1860. After 1880 it was more more commonly used than the Brocot type. The escapement works well with pendulums of between 60 and 180 grammes.

Frequently found in poor quality German and American movements is the type of anchor recoil escapement shown in Fig 4.10, which consists of a bent metal strip.

Fig 4.9 The English anchor escapement.   *Fig 4.10 The inferior bent strip escapement.*

The Brocot Pin-Pallet Escapement

The popular Brocot pin-pallet escapement in Fig 4.11, is a form of dead beat escapement that was patented by Louis Gabriel Brocot in 1841. Louis Brocot worked in Paris from 1820 until 1850; his son Achille Brocot (1817-1874) made several notable inventions relating to clocks. 'Dead beat' means that the escapement does not recoil.

The escapement is simple to make and is ideal for domestic clocks. There are, however, technical objections to it being used in more sophisticated timepieces. English clockmakers objected to the design, saying that lubrication of the pallets was a problem, especially when steel pallets were used. Pallets were often jewelled, using a jewel stone such as cornelian, thereby reducing the lubrication problem. In my experience the escapement works well unless it has been 'attended to' by an inexperienced worker.

72

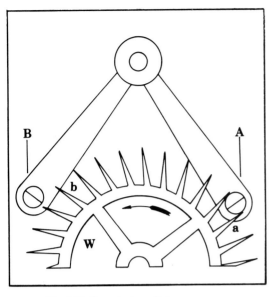

Fig 4.11 Brocot pin-pallet
escapement viewed from rear of
movement.

The escape wheel W will move anti-clockwise under the power of the mainspring and going train. When the pendulum swings to the left, pallet A is swung aside to release the tooth a. The escape wheel jumps a short distance and is stopped by tooth b striking the curved side of pallet B. The pendulum swing a little more, comes to a stop at the end of its swing to the left and, when it reverses, is impulsed by the tip of tooth b pressing on the curved face of pallet B, which pushes the pallet out and upwards. The shape of the semi-circular pallets ensure that there is no recoil because each pallet slides a little way down the straight side of a wheel tooth at the ends of the pendulum's swing.

The Brocot Visible Escapement

Fig 4.12 Brocot
visible escapement.

The visible version of the Brocot escapement (Fig 4.12) is undoubtedly the best known because it can be seen on the front of the clock dial. Fitted with a highly polished escape wheel and fittings, it is an attractive feature for any marble clock. The pallets are normally made of a hard stone such as agate. It was developed by Brocot's son, Achille, from his father's original escapement. This later version was fitted with larger pallets.

The operation of the escapement is identical to that of the first Brocot model. When viewed from the front, the escape wheels on both types of escapement turn clockwise. The exception is when a seconds hand is fitted, in which case they run anti-clockwise.

If new pallet pins are being made for either of the Brocot escapements, ensure that the finished diameter of the rod employed is equal to the distance between the teeth of the escape wheel. It may be very slightly smaller but certainly not greater and all measurements should be made with a micrometer. The pin faces

must be cut to exactly half the diameter of the rods, anything less than half will cause a loss of impulse, and any more could cause the pins to jam between the teeth of the escape wheel.

Brocot Escapement with Two Wheels

The unusual Brocot escapement with two wheels in Fig 4.13, was invented by Louis-Gabriel Brocot and exhibited at the Paris Exhibition of 1839. It was designed so that it could be made into a recoil escapement or a dead-beat type by merely adjusting the height of the pallet.

The two visible wheels are of identical diameter and number of teeth. Each is connected to an identical wheel within the movement plates which forms part of the going train. The escapement has just one pallet that may be raised or lowered without changing the centre of its pivot. The escapement becomes more dead-beat type as the pallet is raised, and more of a recoil type as the arm is lowered. The clock shown in Fig 1 of Collectors' Clocks at the end of the book shows one in working order.

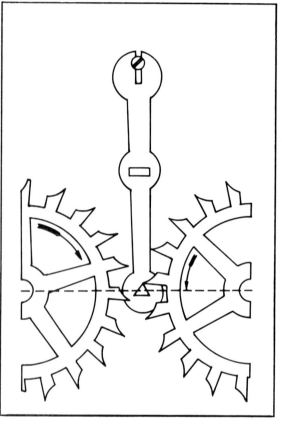

Fig 4.13 A Brocot escapement with two escape wheels.

Adjusting Pallets

In order to adjust the freedom of the escapement pallets, the movement has a device (Fig 4/14, on facing page) enabling the pivot hole to be moved. The pivot hole is placed eccentrically in what amounts to a small turntable. Rotating the turntable, using a screw driver in the slot provided, allows the inside and outside freedom of the pallets to be adjusted. However, adjustment will usually be found unnecessary. If it is, first refer to one of the advanced books noted in the bibliography at the end of this book, but be careful because this is one of the tasks beyond the skill of a beginner.

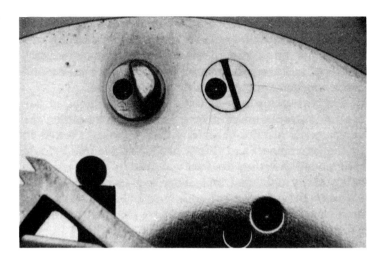

Fig 4.14 The front pivot hole of the anchor (top right) is positioned eccentrically in a bush in the plate. This forms a miniature turntable that can be rotated by a screwdriver, placed in the slot, to adjust the depthing of the pallets. The left hand turntable is moved with pliers. (The illustration is from a striking movement). The pivot holes are in the turntables and the two below have oil sinks.

The Lever Escapement

The lever escapement, Fig 4.15, was invented in 1754 by the celebrated horologist Thomas Mudge (1715-1794). It was first used in a watch made for Queen Charlotte in 1769. Born in Exeter, England, Mudge was apprenticed to George Graham, successor to Thomas Tompion. He devoted his later life to producing an accurate marine chronometer. In French marble clocks, the lever

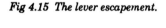

Fig 4.15 The lever escapement.

escapement will occasionally be found as a separate unit incorporating a wheel balance and spring and fitted to a base, when it is known as a 'platform escapement'. Platform escapements dispense with the pendulum and make a movement considerably more portable.

The extensive use of hard stones on most of the bearing surfaces make it long lasting and reliable. The escapements are not however suited to being worked on by the untrained hand. From bitter experience, I would recommend that they are best repaired by a professional, being easily damaged and expensive to replace. If the escapement is original and has been damaged by unskilled hands, its replacement by an identical model will only reduce the value of the clock as a whole.

Oil Sinks

Oil sinks are very small and simple, but are vital parts of the clock plates. The shafts or pivots of the wheels run in pivot holes drilled in the plates. The holes are countersunk on the outside of the plates in order that they may retain a small drop of oil. The ends of the pivots in the oil sinks can be seen in Fig 4.14 and the drawing in Fig 4.16 illustrates the principle of holding a spot of oil in the desired place. Many early French movements did not have oil sinks. The idea of the sink was developed during the early part of the eighteenth century by the Englishman Henry Sully (1680-1728). At one stage in his career he worked in France, where he was befriended by the celebrated French horologist, Pierre Le Roy, who subsequently perfected the oil sink idea.

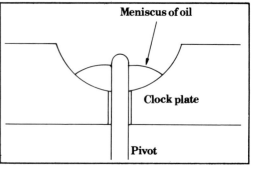

Sectioned view of an oil sink showing the meniscus of oil.

The mid-eighteenth century saw its widespread use in new clocks and many earlier ones were modified. Sinks must not be overfilled with oil; if they are, there will be a tendency for the oil to creep away from the sink on to the rest of the plate. They function best when they are relatively deep and narrow in diameter rather than broad and shallow, as long as this is not at the expense of the bearing surface for the pivot.

The Calibres of Movements

The styles and sizes of clock and watch movements are referred to in the horological world as 'calibres'. To most people the word 'calibre' suggests guns and rifles, but in this case it is used to describe not only the dimensions of the plates, but also the overall style and character of the movement. Here we are only interested in the dimensions of the plates.

L'Industriel Calibre

The diameter of most movements is approximately 81 mm, or to use the traditional French measurement, 3 pouces. A pouce is a French inch equal to 27.07 mm. In France such movements are referred to as 'Les Mouvements Industriel'. They appeared in 1810 and, until 1850, usually had an English anchor escapement together with a silk suspension (see Fig 3.15). After 1850, the spring suspension was gradually introduced.

The Mignonnette Calibre

'Mignonnette' is not readily translated into English; I would call it the 'petite calibre' using the English word denoting something that is small and dainty. The movement is so named because it is the smallest of the pendule de Paris type, measuring 2³/4 pouces or 74 mm. It is quite rare and usually found only in smaller cases. Unfortunately it is not possible to date it.

During the period 1790-1830, a calibre of 4 pouces (108 mm) was popular.

The last calibre measures 3¹/3 pouces (90.5 mm) and was built from 1880 through to around 1930. It was usually used in top quality clocks and normally had rack striking and a Brocot type of escapement.

The Hands

With antique clocks of many kinds and ages, the movements can be dated by their hands, if these are original (Fig 4.17). Distinct changes in style and construction will usually enable the specialist to give a reasonably accurate age to the movement.

Unfortunately French marble clocks had only had four basic shapes of hand and these were imitated from various earlier periods of French horological style. The shapes used were popular during the 19th century and their use was not restricted solely to France. Split Moon

77

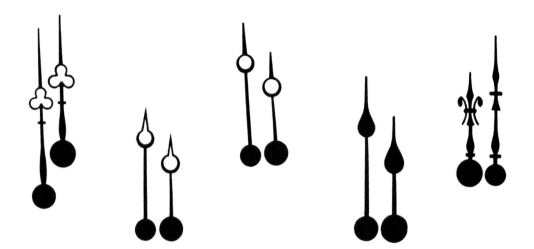

and trefoil hands are the only types - again if they were the original hands - by which a French marble clock may be attributed to an earlier period.

Fig 4.17 Styles of hands (L to R): (1) The trefoil and (2) Split Moon shapes, c. 1840 - 1850, (3) Moon and (4) Spade, from c. 1855 until 1910, (5) Fleur de lys, c. 1885 - 1910.

V

Understanding the Pendulum

At first sight the humble pendulum (Fig 5.1) appears to be one of the simplest parts of a clock; unfortunately this is far from true. It plays a crucial and often underestimated role in the regulation of a movement, using the force of gravity to provide a regular 'beat' and force the movement to release the power stored in its spring very slowly and regularly.

Controversy has always surrounded the identity of its inventor. At one time or another, Leonardo da Vinci, Galileo Galilei, and Christiaan Huygens have been given the credit, although the consensus of opinion now is that both Leonardo da Vinci (1452-1519) and Galileo Galilei (1564-1642) evolved a theoretical concept, while it was left to the Dutchman Christiaan Huygens (1629-1696) to develop the first practical application of the pendulum in 1656.

At the the age of 19, Galileo noted that the times taken by a pendulum to swing from one side to another appeared to be equal. It was later established by Huygens that the swings are nearly equal for short swings only; in longer swings there is sufficient difference to cause problems. Galileo conceived his theory after watching a lantern swinging from a long chord. He timed the oscillations with his pulse; a technique also used by ancient mariners to ascertain the speed of a ship. (They timed the passage of a piece of

Fig 5.1 The pendulum with figures (a) on the back plate giving its length in pouces and lignes.

wood from the stem of a ship to its stern, which is how the nautical term 'log' originated).

It is not possible to make a working model of a theoretical pendulum, but an approximation could be made by suspending a small ball from a fine thread and making it swing. If the experiment were conducted in Greenwich, England, the ball would take one second to swing from one side to another and be known to clockmakers as a 'seconds pendulum', as long as the distance from the suspension point to the middle of the ball was 0.99413 metres (39.1397 inches). The famous London clock, 'Big Ben', has a pendulum with an effective length of about four metres or thirteen feet. This gives a two-second beat, so that the pendulum swings from one side to the other in two seconds and the minute hand moves every two seconds.

The usual formula for relating the length of a pendulum to the time it takes to oscillate is π times the square root of the length of the pendulum divided by the force of gravity. That is:-

$$t = \pi \sqrt{L/g}$$

$t = \pi$ times the square root of L/g where
t = time of swing in seconds,
L = length of pendulum, and g = force of gravity.

From the formula it is possible to see that changes in gravity directly affect the timing of the pendulum. The force of gravity varies according to your position on the Earth's surface. For this reason, a pendulum beating seconds at the Equator would be fractionally shorter than one at the North or South Pole.

Calculating the Length of a Pendulum

For practical purposes, g can be ignored when calculating pendulum lengths for domestic clocks. Moreover, if the pendulum from a clock is lost it will not usually be necessary to calculate the length of a replacement pendulum.

If the movement is in a case, the simplest method would be to buy the longest pendulum that fits the case; the timekeeping of the movement is then monitored and the pendulum gradually shortened until a degree of accuracy is obtained. Alternatively, one of the following methods could be used.

The French were very professional in their approach

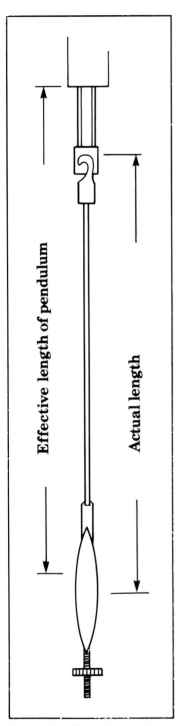

Effective length of pendulum

Actual length

Fig 5.2 The effective length of a pendulum.

to the manufacture of their drum movements. Apart from the excellent quality of production, they took great care in the general design, giving attention to small but apparently important details.

Most back plates will have markings on them that indicate the effective length of their associated pendulums. On the bottom of the plate there will often be two numbers separated by the lowest pillar. Looking at the movement in Fig 5.1, the letter a points to the figures *4* and *11* stamped on the bottom of the back plate. This indicates the effective length of pendulum required. Not all movements will have such a mark as it did not come into regular use until after about 1860.

The left hand number *(4)* is in pouces and the right hand one *(11)* in lignes, giving the effective length of the pendulum as 4 pouces 11 lignes. A pouce is a French inch (1.066 English inches or 27.07 mm), and a ligne is a twelfth of a French inch (0.088 English inches or 2.256 mm), so the length in millimetres is (4 x 27.07) + (11 x 2.256) = 133 millimetres for practical purposes.

Measurement of the pendulum that came with this movement showed that its length was correct. Always remember that the effective length of the pendulum is not its overall length, but its length from the centre of suspension to the centre of oscillation (Fig 5.2).

Another Method

The method suggested above is used when the pendulum length is marked on the back plate of the movement. Without this, it is necessary to employ a tedious calculation that involves counting the individual teeth of the wheels in the going train. The major drawback of this method is that the movement will have to be taken apart to count the teeth of the various wheels. This is the procedure:-

Count the number of:-

[a] Teeth on the centre wheel
[b] Teeth on the third wheel
[c] Teeth on the escape wheel
[d] Leaves on third wheel pinion
[e] Leaves on the escape pinion

81

The number of oscillations the pendulum will make per hour is (a.b.c)/(d.e) so that for the clock in question the figures are:- (84 x 70 x 41)/(7 x 7) = 4920. The pendulum will therefore make 4920 oscillations an hour.

The next step is to square 4920 and to divide the result into the constant 320,000,000. This is a simplification of the general equation for a simple pendulum. So the length is 320,000,000/(4920 x 4920).

Dividing through by 100 simplifies it to 320,000/(49.2 x 49.2) = 132 mm for practical purposes.

Types of Pendulum Encountered

Several different types of pendulum are found in the movements of French marble clocks. They range from the simple brass bob to the elaborate and impressive mercury compensated type. Most will have pendulum bobs similar to those illustrated in Fig 5.3.

Fig 5.3 The simpler pendulums, from the left: Simple brass shell filled with lead; Brocot pendulum; Thieble pendulum (note the absence of a squared shoulder above the bob); Plain solid bob with set screw holding it in position on the rod.

An exception is a pendulum with a hook (Fig 5.4), designed to be used with silk suspension. These pendulums are adjustable by a knurled nut at the bottom which enables the pendulum bob to be raised or lowered on its rod. Raising the bob will increase the speed at which the pendulum swings, which in turn makes the clock run faster. Similarly, moving the bob down will slow the movement.

The fourth type has a set screw that holds the

Fig 5.4 A silk suspension. The pendulum hangs from a thread of silk instead of a steel spring. (A white card has been positioned to aid viewing). The main regulation is by moving the bob; fine regulation is from the top by winding up the silk thread.

pendulum in position against the rod, also allowing simple yet effective adjustment of the bob's position, although not as convenient as a thumb screw.

Compensating Pendulums

A well-known principle in physics is that a metal will expand when it is heated and contract when it is cooled. This is true of metal pendulum rods. Great accuracy is not required of normal domestic clocks. The effect was often ignored by manufacturers, but is taken into account in more accurate clocks, and in special precision timepieces, such as those called 'regulators'.

If a clock is fitted with an ordinary pendulum, any increase in its length will tend to make it beat slower and any shortening will increase the rate. For this reason much research was undertaken in the early days to find a method of compensating for the variations in length caused by fluctuations in temperature. Two types of compensated pendulum are usually found in French marble clocks, the Ellicott and the mercurial.

The Ellicott Pendulum

The pendulum shown on the next page in Fig 5.5 was devised by John Ellicott (1706-1772), the son of a London clockmaker. He became a celebrated horologist, making both watches and clocks of fine quality. His design for the compensated pendulum was originally revealed in a paper given to the Royal Society in 1752. The cantilever principle was employed, but the device resulted in a considerable degree of friction, and the pendulum offered only partial compensation. It is said to have been discarded in favour of the mercurial pendulum. Despite this, it is frequently found in good quality 19th century French mantle clocks and is occasionally found in French marble clocks.

The Mercury Pendulum

Fig 5.5 Compensating pendulums, the Ellicott on the left and mercurial on the right.

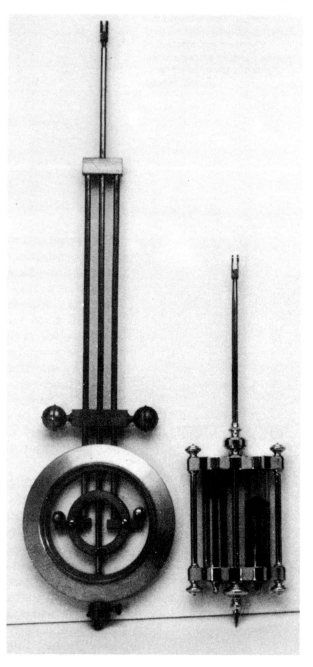

The mercury or mercurial pendulum, on the right in Fig 5.5, was developed by the famous English clock and instrument maker George Graham. Graham discovered that a cylindrical jar of certain proportions, when filled with a certain quantity of mercury, would compensate for the downward expansion of a pendulum in heat. Glass jars are usually used, although steel or iron jars are preferable as they have the ability to react more quickly to temperature changes than glass.

French makers favoured the two-jar version as modified by the Frenchman, Duchemin. The arrangement involves more labour and care in its making but has the advantage of displacing the centre of gravity of the liquid less. It also responds quickly to changes in temperature.

When fitted to ordinary marble clocks, both mercury and Ellicott pendulums, with rare exceptions, can be ignored as far as their affect on timekeeping is concerned. They were only used for decorative purposes, and are certainly not functional.

The pendulum can provide invaluable information helping to date a movement, as outlined in Chapter Three. One point should be made about the inscriptions often found on the pendulum bob. Firstly, you should check that the serial number stamped on it is identical to that on the rear plate of the movement. Secondly, much information can be gleaned from the whispy scratched handwriting of previous restorers.

Weights of Pendulum Bobs

When a clock is bought, check that the serial number stamped on the pendulum bob is identical to that on the rear plate of the movement. The weight of a correct pendulum bob depends on the type of escapement used, as follows:-

(a) Small recoil (anchor) with silk suspension . . 30 grams
    Small recoil with spring suspension . . . . 30-60 grams
(b) Gable anchor with spring suspension . 100-500 grams
(c) English anchor with spring suspension . . . . .70 grams

(d) Brocot with spring suspension . . . . 100-500 grams
    and with mercury pendulum . . . . 250 grams

These are nominal values and may vary from one movement to another.

The Suspension

The suspension is one of the most important parts of the clock. Any small kinks in the suspension spring will cause difficulties with either running or timekeeping. Similarly, a spring that is excessively strong or weak will affect the timekeeping. One of the most common faults encountered with French marble clocks is a spring that has been accidentally twisted or broken. Usually this has happened when someone who is ignorant of how a clock works has moved it with the pendulum still attached to the spring.

Surprisingly, this fault can be a bonus for buyers of unrepaired marble clocks, which are often sold cheaply because they do not.work. Not infrequently, the only fault is the suspension spring and a new one costing a few pence will bring a discarded clock into life again.

Christiaan Huygens used a silk suspension similar to that in Fig 5.4 with his first pendulum in 1657. It was widely employed in French movements until Louis-Gabriel Brocot developed his popular arrangement universally known as the 'Brocot suspension'. Although introduced in 1840, it is not widely seen in clocks dated before about 1860.

85

The first experimenter of the Royal Society, Dr Robert Hooke (1653-1703), had the idea of replacing the silk suspension with a piece of flat spring. The device was subsequently demonstrated to members in 1666. Hooke also discovered the law of the spring, that the power stored in a spring is related directly to its tension. William Clement (1638-1704) has also been credited with devising the suspension spring.

The invention was improved in 1710 by the Frenchman Julien le Roy (1686-1759), who used two springs in the form still seen today. Three examples are shown in Fig 5.6. This reduced unwanted wobbling of the pendulum. The arrangement was used only in quality clocks until the arrival of Brocot's suspension in 1840.

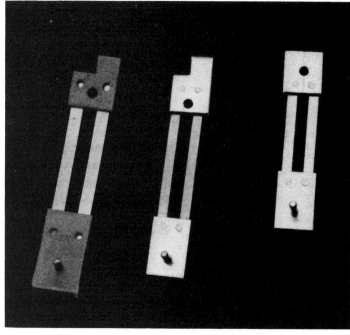

Fig 5.6 Suspension springs, left to right; Brocot pattern with modern plastic ends: Brocot with brass ends; Vallet type with brass ends.

The Brocot Suspension

Brocot's invention was widely implemented by French makers. The Brocot family produced many useful ideas and products that played a considerable part in advancing the progress of the French movement, unfortunately there is little information available about them apart from the usual patent dates and entries in horological dictionaries. The most information that I have found is in a delightful little book entitled the 'New and Complete Clock and Watchmakers Manual' (New York 1877) by an American lady called Mary L Booth. From this work some interesting ideas arise about Brocot's inventions and work.

In the first place, many writers give the impression that Brocot invented the suspension around the year 1841, but had to wait several years until the 1850s before it became popular. Booth's book includes details of Brocot patents dating from October 9, 1840. The first is entitled: 'Patent of invention for five years, for improved movements of clock-work, by M. Brocot, of Paris; dated October 9, 1840'. An interesting point is that the patent was granted for five years. Today we think of patents in terms of years or even decades. It is likely that Brocot's ideas were not taken up by the rest of the industry until the patents had expired.

Subsequently other amendments were made to the

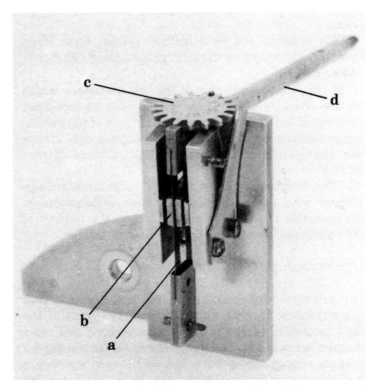

initial patent. The first patent for addition and improvement was dated November 14, 1840, and the second, June 20, 1842. The latter would appear to be the one with which collectors are familiar (Fig 5.7). The details of this patent mention a double spring for the suspension, and the design of the cheeks of the suspension block are:-

'A suspension with a double band is also used. In the new arrangement the clasp has three clefts; that of the middle still receives the suspension spring, while the two others, made nearly at the edge, form by drawing them a little aside, two springs which press on the inner cheeks of the frame and produce a good and indestructible adjustment'. While it would be easy to gain the impression that Brocot's invention was revolutionary, a

87

wiser approach would be to look upon it as the synthesis of several previous ideas. The celebrated English maker Thomas Tompion (1639-1673) had used an essentially similar idea in his bracket clocks. An early French cartel clock with the name 'Revel Paris' (1775-1789) has been seen with a suspension that predates Brocot, and bears a marked resemblance to the suspension given his name. Similar examples have been found on other early French clocks.

This is borne out by a further extract from Mary Booth's book regarding the first patent dated October 9, 1840:-

'M. Brocot, the inventor of the improvements which we are about to describe, perceived that he had been anticipated in the discovery of the principle of pendulum-compensations by M. Wagner, he therefore only claims the application of the material conditions of this principle'.

The Wagner mentioned here is probably Jean Wagner, who, amongst other activities, collaborated in the invention of the metronome. He was the nephew of the celebrated horologist J Bernard Henri Wagner.

Operation of the Suspension

The principal idea embodied in the suspension is that the pendulum may be adjusted from the front of the clock instead of having to turn the case round and move the pendulum bob up and down. It can be seen in Fig 5.7 that the suspension spring a is held between the cheeks b of the suspension block; the cheeks do not pinch the spring but merely hold it in close contact.

The adjusting wheel c is turned by the rod d that projects through the top of the clock's dial. The wheel, acting on a vertical screw thread, causes the suspension block cheeks to be raised or lowered depending on the direction in which the wheel is turned. Raising the block will increase the effective length of the pendulum and consequently slow the rate of the clock; turning the wheel the opposite way will cause the clock to gain time, ie run faster.

The Vallet Suspension

The suspension attributed to Vallet in Fig 5.8 is similar to Brocot's but has the advantage that the moving parts are totally enclosed. It is frequently found in the movements of French marble clocks. The main points to watch when working with it are that Brocot and Vallet springs are not interchangeable. From Fig 5.7, it may be seen that the Brocot type has a projection of brass from its upper edge. From a practical point of view, the Vallet type of suspension is easier to strip and clean than the Brocot because it has fewer parts.

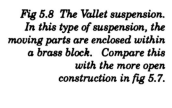

Fig 5.8 The Vallet suspension. In this type of suspension, the moving parts are enclosed within a brass block. Compare this with the more open construction in fig 5.7.

The Duvoye Suspension

The Duvoye suspension is in many respects a cross between a Brocot and a Vallet versions. The name 'Duvoye' appears on the adjusting wheel. I first saw it used on a Japy locking plate movement that was manufactured after 1880. Duvoye lodged a patent for a type of clock suspension in 1852.

VI

Rack and Locking Plate Striking Movements

French marble clock movements are among the smallest and most complex to be found in the world of domestic clocks. Some writers have likened their fineness and complexity to that of large watches. It is no easy task to explain their operation to someone ignorant of clocks by trying to relate two-dimensional pictures in a book to the three-dimensional mechanism that sits menacingly on the work bench.

It is unfortunate that some of the technical terms do not lend themselves to an easy acquaintance, but every effort should be made to learn the names of the various parts of the movement. For the purposes of this chapter I have assumed that the reader has never seen, or even handled, a striking movement previously. The explanation here about the rack striking movement (Fig 6.1 and 6.2) is intended to be detailed; far too many books gloss over the operation of movements, leaving too much to the reader's imagination.

Fig 6.1 The back plate of a rack striking movement with the bell removed.

Fig 6.2 *Under the dial view of a striking movement showing the front plate and rack striking mechanism mounted on it.*

How Rack Striking Works

The main components and their functions are explained in the next paragraphs, which refer to Fig 6.3 to Fig 6.5. More special names for special parts are introduced in this chapter and, although many will be strange, it is important to become familiar with them because they are universal in horology (the science and practice of time measurement).

The barrel and spring (*a* in Fig 6.3) of this train are very similar to those in the going train. The sole purpose of the spring here is to provide a power source for driving the striking train and hammer. For the purposes of understanding the strike action its operation may be ignored.

The intermediate wheel (*b* in Fig 6.3) has no direct effect on the starting and stopping of the train; it is merely a form of idler wheel. It too may be ignored for the moment.

Fig 6.3 Rear view of a rack striking movement on which the brass back plate has been replaced with one made of clear Perspex: (a) Strike barrel and main wheel containing the driving spring, (b) Intermediate wheel, (c) Hammer arbor with its squared end, (d) Pin wheel, (e) Strike stop arm, (f) Gathering pallet wheel, (g) Warning wheel, (h) Fly.

The pin wheel (*d* in Fig 6.3) is so named because it has many pins on it. This wheel lifts the hammer that in turn strikes against the clock's gong or bell. For the rest of this chapter the hammer will be described as hitting a bell; gongs are ignored for the present. Looking at the movement from the front, assume that the strike train is running; the pin wheel will be driven by the train in a clockwise direction. The pins on the wheel lift up a steel peg that projects from the hammer arbor *c*. The rear end of the hammer arbor has the hammer squared on to it. As the pin wheel revolves, the hammer is lifted and drops back against the bell. A long thin spring, the hammer spring (not visible in the picture), helps accelerate the hammer head towards the bell.

How the bell is struck is quite simple; how the strike train is started and stopped is more complicated.

The cannon pinion (Fig 6.4) is located on the arbor of the centre wheel. The minute hand is fixed to the front end of the cannon pinion, which has a square section of brass allowing the hand to be precisely positioned. (The squared end can be seen in the centre of Fig 6.5). Remember that the motion of the cannon pinion is

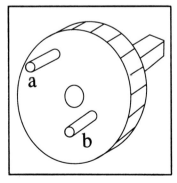

Fig 6.4 The cannon pinion. Notice that the pins are at an unequal distance from the centre of the wheel.

93

directly related to the minute hand's movement around the dial. Projecting from the back of the cannon pinion are two small steel pins, one of them is nearer to the centre of the wheel than the other, as can be seen from the illustration. The inner pin initiates the half-hour striking, while the outer pin looks after the hours.

With the clock running normally, each pin will eventually come into contact with the lower edge of the long blade of the lifting piece *j* in Fig 6.5, causing it to be moved upwards. For the moment note that the outer or hour pin will move the lifting piece further than the inner or half-hour pin.

Lifting piece, warning lever, and rack hook: In Fig 6.5 it will be seen that above the lifting piece there is another arm, the warning lever *k*, the two forming a V on its side. These two arms are fixed together and therefore work together, so that when the lifting piece *j* is moved upwards by one of the pins on the cannon pinion, the warning lever is raised too.

Fixed to the inner side of the S-shaped arm of the rack hook *l*, is a pin, shown dotted at *m*, which rests on the upper edge of the lifting piece *j*. The rack hook is squared on to an arbor mounted between the plates of the movement which also carries a lever called the 'strike stop arm'.

The strike stop arm (Fig 6.3 *e*) is a short arm fixed at right angles to the rack hook arbor and moves with the rack hook. Its function is to stop the strike train when it is running. The widened end of it, called a 'nose', can move into the path of a solitary pin on the edge of the gathering pallet wheel *f* in Fig 6.3.

The Warning Cycle

The purpose of the warning cycle is to make the striking train start striking the bell or gong at an exact moment. When the minute hand approaches the hour, the outer (hour) pin on the cannon pinion (Fig 6.4) will start to raise the lifting piece (Fig 6.5 *j*). The lifting piece in turn raises the strike stop arm (Fig 6.3 *e*), until its nose is clear of the pin on the gathering pallet wheel (Fig 6.3 *f*). This wheel will then be free and the striking train begins to run. It engages with another smaller wheel, the warning wheel (Fig 6.3 *g*), which also has a pin on it. The

Fig 6.5 The rack striking mechanism: (j) Lifting piece, (k) Warning lever, (l) Rack hook, (m) Rack hook pin, (n) Gathering pallet, (o) Nose on warning lever, (p) Rack, (q) Rack tail, (r) Snail.

train, which has just started to run, is held up again, this time by the warning lever (Fig 6.5 k). The dotted rectangle marked o in Fig 6.5 shows the end of part of the warning lever which projects at right angles through a slot in the plate so that it can intercept the pin on the warning wheel.

To summarise the warning cycle: The lifting piece (Fig 6.5 j) is slowly lifted. This lifts the rack hook l. The strike stop arm inside the movement (e in Fig 6.3) releases the pin on the gathering pallet wheel f, which then turns the warning wheel g in Fig 6.3. The warning wheel should, especially in gong movements, make no more than a quarter turn before being arrested itself by the nose o (in Fig 6.5 this time) of the warning lever k.

The train is now ready to run and strike when it is released at the exact moment on the hour or half hour. This happens when the lifting piece drops off the hour pin of the cannon pinion just as the hour hand reaches

95

the hour. The warning lever drops with it and the stop or nose *o* releases the warning wheel, allowing the train to run on and strike the hours.

Striking the Hours

The correct striking of the number of hours depends on another function, under the control of two main components, a cam shaped like a snail's shell and a rack like a reaper's hook.

The rack, rack tail, and hour snail: The rack (Fig 6.5 *p*) is bonded to its associated component, the L-shaped tail *q*. These two levers therefore pivot together around a pin that projects from the front plate. Under the rack tail is an eccentric cam (Fig 6.5 *r*), the hour snail, which is attached to the hour wheel so that it moves round at the same rate as the hour hand of the clock. Whenever the hour hand is at a particular position, for example at twelve o'clock, the snail will be in the same relative position.

In Fig 6.5, the rack *p* is shown in the position it occupies when the clock is not striking. That is to say, it is in its highest position, held up there by the rack hook *l* underneath its end.

When the hour pin, the outer one on the cannon pinion, lifts the lifting piece sufficiently high, the rack hook *l* will also be lifted to initiate the warning cycle, as explained above. As the rack hook is lifted still further by the lifting piece, the rack hook will move from under the rack, allowing the rack to drop downwards under its own weight. Its fall will be checked by the rack tail *q* hitting the edge of the hour snail. A few moments later, the lifting piece will be free of the hour pin and drop as well, to release the striking train and start the strike sequence.

The gathering pallet wheel: The strike train is now running. One of the train wheels is the gathering pallet wheel (Fig 6.3 *f*); on the dial end of this is mounted a small cam, the gathering pallet n in Fig 6.5. As the gathering pallet wheel turns, the pallet lifts the rack upwards one tooth at a time until the rack is at the end of its travel upwards.

Each time a tooth of the rack is lifted upwards, the bottom right corner of the rack hook is moved from between two teeth of the rack and drops back between the next two adjacent teeth of the rack, preventing the rack from falling back again. When the rack reaches its highest position, and there are no more teeth for the rack hook to drop between, it falls below the rack to hold it in its highest position. When the rack hook drops below the rack, the strike stop arm attached to its arbor (*e* in Fig 6.3) halts the train and secures the rack.

We are now back to the position of the mechanism before striking, as shown in Fig 6.5.

Counting the Hours

A close look at both the hour snail and the rack tail, will indicate that the distance the rack falls is proportional to the height of the cam on the snail. At twelve o'clock the tail falls on to the lowest part; for one o'clock it falls on the highest part. Bear in mind that when the rack drops, the strike train starts to run. It should now be clear that twelve strikes will be produced by the time taken for the rack to be completely lifted or gathered up over the long distance from the lowest point on the snail. Conversely, at one o'clock only a short distance is involved and the strike train only has time to strike the bell once.

Striking the Half Hours

When the movement nears the half past position, the pin nearer the middle of the cannon pinion wheel will raise the lifting piece slightly, thereby initiating the warning cycle. It does not, however, lift the piece far enough to push the rack hook clear of the rack. As the pin continues to move, the lifting piece drops and the strike train starts. This time the gathering pallet wheel, and the pallet attached to it, make just one revolution before being checked by the strike stop arm.

There are a few other components in the strike train that have not been mentioned.

The fly (Fig 6.3 *h*) is the last element of the striking train, it is a fan or air-brake acting as a form of governor and regulates the speed of the train, affecting the rate at which the bell strikes. The three components of the fly are a steel arbor, a thin piece of brass sheet, and a small flat spring. The spring is designed just to grip the blade on the arbor. This will allow the blade to slip when the strike train stops suddenly; for this reason a small drop of oil should always be placed on the connection point between the spring and its arbor. If the fly were stopped abruptly, the train might be damaged.

The fly turntable: The depthing of the fly is accomplished by adjusting the turntable in which its front pivot runs (see Fig 4.14). The function of this device is identical to that of the pallet arbor and will not normally need adjusting.

The hammer stop: The hammer is checked on its downward path by a steel pin mounted in the back plate of the movement; *a* in Fig 6.6 indicates where it is located as it is virtually impossible to see in the picture. It sits right under the base of the hammer arm. The pin inevitably gets bent to a suitable position; if a broken one is encountered, remember it is an ordinary tapered steel pin pushed in from the inside surface of the back plate.

The hammer: The hammer head *b* (Fig 6.6) usually consists of a turned piece of brass; on gong movements the head will consist of a hammer held by a screw with a leather button inserted into it. The tone produced can be altered by singeing the leather, or by making cuts across the striking surface with a sharp knife and oiling it.

The head is mounted on a long thin stem made of brass threaded into the boss at its base, this in turn being squared on to the hammer arbor. A common problem occurs when the stem falls out of its hole. Careful use of soft solder will make an acceptable repair, but do not leave a large blob of solder on the joint. Unfortunately it is common for this to have been done at some stage during the life of these movements. The tone produced by the hammer can be adjusted by carefully bending the stem. Hold the lower part of the stem with a pair of long-nosed pliers while doing this or you will weaken the joint between the hammer and the boss.

The pin wheel cock: (Fig 6.6 *c*) In these movements a cock consists of a pivot hole that is sunk into a brass plate. The whole plate is removable from the back plate of the movement; in the case of a striking movement this allows the pin wheel to be set up during the rebuilding of a movement.

Fig 6.6 Back plate of a rack striking movement:
(a) Hammer stop, hardly visible under the base of the hammer arm,
(b) Hammer, (c) Pin wheel cock.

All the components of a rack striking clock are shown in Fig 6.7.

The biggest single advantage of the rack movement over the locking plate variety, is that the strike train is always synchronised with the going train so that the striking is always in step with the hands. The only real disadvantage is that the movement is relatively complex and possibly dearer to manufacture.

A Note for Owners and Restorers

The rack striking movement has a quirk; if the striking train jams or runs down, it will be impossible to lift the rack when it gets to the twelve o'clock position on the hour snail. The going train will be unaware of this problem and the rack tail will eventually collide with the long vertical upright part of the snail, causing the whole movement to stop at or near the twelve o'clock position. Many a French marble clock has been bought cheaply

Fig 6.7 The components of a rack stricking movement:
(1) Back plate, (2) Front plate, (3) Going barrel, (4) Strike barrel, (5) Escapement and pendulum crutch, (6) Pendulum suspension block, (7) Escape wheel, (8) Intermediate wheel, (9) Third wheel, (10) Centre wheel, (11) Warning wheel, (12) Gathering pallet wheel, (13) Pin wheel, (14) Intermediate wheel (strike train), (15) Hour wheel and hour snail, (16) Minute wheel, (17) Cannon pinion, (18) Minute wheel cock, (19) Fly, (20) Rack hook, (21) Strike stop arm and arbor of rack hook, (22) Spring for 21, (23) Lifting piece and warning lever, (24) Rack and rack hook, (25) Bell stand, (26) Hammer, (27) Hammer arbor, (28) Spring for 27, (29) Ratchet wheels, 30) Ratchet wheel cocks, (31) Clicks, (32) Click springs.

100

when this was the only problem in a clock that was sold 'as seen'.

Some racks have a bevelled tail that enables the snail to push the rack upwards and out of the way should the strike train have failed to function.

Fig 6.8 *Back plate of a locking plate striking movement, showing the locking plate and the knife edge resting on it. (Bell removed).*

How Locking Plate Striking Works

Locking plate striking movements (Fig 6.8 and 6.9) are considered to be easier to work on than rack striking movements, but this does not prevent beginners from occasionally having with problems setting them up. It is important that the theory behind the striking is understood because, when this is mastered, the work becomes a matter of routine.

A cursory glance at Fig 6.9, will show that there are fewer parts to be found on the front plate of the locking plate or count wheel movement, when compared with the rack movement. The snail and rack increased the production cost of the movement, which is why locking plate movements were sometimes used to reduce manufacturing costs.

The cannon pinion, *a* in Fig 6.9, is a friction fit on the centre arbor. On the rear side of the cannon pinion are two pins (Fig 6.10). These pins are mounted opposite one another and are equidistant from the centre of the pinion, unlike those in the rack movement. As the circular path of both pins is synchronised with the minute hand, one pin (the hour pin) is at the top at the hour and the other at the half hour.

101

Fig 6.9 Front plate of a locking plate striking movement: (a) Cannon pinion, (b) Lifting piece, (c) Warning lever, (d) Strike stop wheel, (e) Lever attached to the arbor of the knife edge lifting lever shown at (e) in Fig 6.11, which can also be seen in this picture.

The knife edge lifting lever: As in the rack mechanism, there is a lifting piece (*b* in Fig 6.9) and a warning lever *c*, fixed together in the shape of a V so that they move together. When the minute hand is approaching the five minutes to the hour position, the hour pin will start to lift the lifting piece and warning lever. The warning lever in turn raises a bow-shaped component, *e* in Fig 6.9, the arbor of which goes through the back plate to near the edge of a large wheel, and carries the knife edge lifting lever itself, as can be seen at *e* in Fig 6.11.

The knife edge (which looks almost black in Fig 6.11) is at right angles about half way along the lever. The top end of this lever engages pin *h* on the strike stop wheel. When the lever is raised, the pin is released, allowing the train to start and initiate the warning cycle, the function of which was explained earlier in the chapter under the heading The Warning Cycle.

The train is soon stopped, however, by a pin on the adjacent warning wheel *i* that meets the brass nose of the warning lever (Fig 6.9 *c*) projecting through a slot in the front plate.

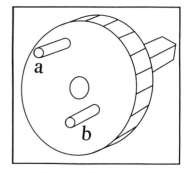

Fig. 6.10 The cannon pinion of a locking plate movement from the rear. The hour and half hour pins that initiate the striking are equidistant from the centre.

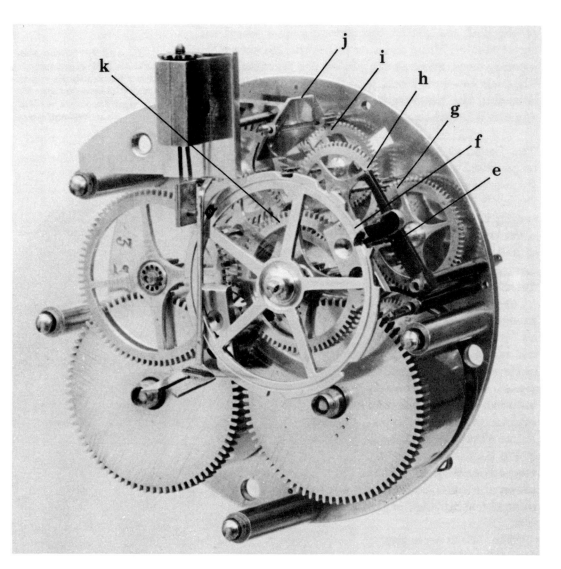

Fig. 6.11 Rear view of a locking plate striking movement. The brass back plate has been replaced with one made of clear Perspex. (e) Knife edge lifting lever, (f) Locking plate, (g) Pin wheel, (h) Pin on strike stop wheel, (i) Warning wheel, (j) Fly, (k) Intermediate wheel.

Striking the Hours

To summarise, the lifting piece and warning lever have been lifted partially by one of the pins on the cannon pinion wheel as it approaches the hour. This has moved the locking plate knife edge sufficently to start the train running. The train has been promptly stopped by the pin on the warning wheel engaging the nose of the warning lever. The warning sequence is now complete.

The cannon pinion wheel will still continue to move round. When the minute hand is pointing to twelve or six on the dial, the pin on the cannon pinion wheel will travel past the lifting piece. As this drops, the nose of the warning lever releases the pin on the warning wheel. The train now runs freely, with the pin wheel lifting and releasing the hammer to strike the hours. All that remains is to find out how the hours are counted.

Fig 6.12 The locking plate with the knife edge resting in the slot between the eleven and twelve o'clock cams after the clock has struck one blow at half past eleven

The locking plate is a large wheel on the rear of the movement. In Fig 6.12, the knife edge can be seen resting in a slot in the edge of the locking plate, which is of brass with a series of cams on its edge formed by the slots.

Assume that the movement has just struck half-past eleven, which is one strike on the bell. The knife edge of the locking plate will be resting between the longest and second to longest cams. Notice that the cams, seen in Fig 6.12, have a slightly ramped leading edge that allows the knife to slide up on to the upper edge of the cam.

When the train begins to run, the locking plate will start to turn in a clockwise direction when viewed from the back. The knife edge will slide up on to the twelve o'clock cam. The length of this cam has been designed to allow the bell to strike twelve times before the knife edge slips off the cam and back into the next slot. When that happens, it causes the arm on its arbor to engage with the pin on the strike stop wheel and stop further striking.

104

Striking the Half Hours

Assume that the movement has just struck eleven o'clock and the knife edge is in between the cams for eleven and twelve o'clock and slightly lower than in the position shown in Fig 6.12. When one of the cannon wheel pins comes round again, it will first of all lift the lifting piece slightly, thereby initiating the warning cycle. When the lifting piece drops, the striking train will start to run. This time however the locking plate knife edge is sitting in a trough between two cams. Because the knife edge is in a trough, its arm will be in the path of the pin on the strike stop wheel. Consequently the wheel will turn once only, which will allow one strike and then stop.

The only part of this sequence that may cause confusion is the one o'clock strike as this is created simply by repeating the half-hour strike pattern. A cam does not therefore exist for one o'clock. The single blows for one, and then half past one, are produced by extending the slot responsible for the single blow at the half hour after twelve o'clock, with the result that three single notes follow each other at half hour intervals for half past twelve, one o'clock, and half past one.

Advantages and Disadvantages

From the point of view of the inexperienced restorer, the locking plate movement has the advantage that there are fewer parts than on the rack movement, making its operation simpler to understand. The main drawback from an operational point of view is that the hand positions are not automatically synchronised with the locking plate position and striking.

It will sometimes be necessary to reach inside the back of the clock case and lift the knife edge, allowing the strike train to run until the correct hour position is found. This simpler movement of earlier design would, however, have made a significant reduction in manufacturers' production costs. All the component parts of the locking plate movement are shown in Fig 6.13.

A Note for the Restorer

Considerable problems will be encountered if the locking plate wheel is removed during restoration and not

105

replaced in exactly the same place on the arbor that it came from. Quite often the square will have a mark on it that corresponds with another mark on the back of the wheel. If this precaution is not taken, the movement could end up by striking in the wrong sequence. Wear in the locking plate arbor bush will produce similar problems.

If the strike train does not lock correctly, it is sometimes worth removing the locking plate and replacing it a quarter of a turn further round before trying again. If this is still unsatisfactory, the pin wheel cock will have to be removed to allow the train wheel to which it is attached to be moved by one or two teeth.

Remember that the strike spring MUST be unwound before removing the pin-wheel cock.

Fig 6.13 The component parts of a locking plate movement: (1) Front plate, (2) Back plate, (3) Going barrel, (4) Strike barrel, (5) Escapement and pendulum crutch, (6) Pendulum suspension block, (7) Escape wheel, (8) Intermediate wheel, (9) Third wheel, (10) Centre wheel, (11) Warning wheel, (12) Strike stop wheel, (13) Pin wheel, (14) Locking plate wheel, (15) Hour wheel, (16) Minute wheel, (17) Cannon pinion, (18) Minute wheel cock, (19) Fly, (20) Lifting piece and warning lever, (21) Locking plate knife edge, (22) Knife edge lever spring, (23) Knife edge lifting lever, (24) Locking plate, (25) Hammer, (26) Bell stand, (27) Hammer arbor, (28) Spring for [27], (29) Ratchet wheels, (30) Ratchet wheel cocks, (31) Clicks, (32) Click springs, (33) Hands.

VII The Restoration of a Timepiece

The restorer of the French marble clock is fortunate because the movements were made to a very high standard and have withstood the test of time well. It seems that most French clocks never wear out because they are so well made; the pinions are hard, and repairs needed to the movement as a whole are generally modest in extent. Unfortunately, the very fineness of the component parts can pose hazards to either an inexperienced repairer, or one who is more familiar with larger, robust pieces. Great care should be exercised when unpinning and pulling the plates apart; some pivots are extremely thin and may snap if roughly handled.

Fig 7.1 Front and rear views of a timepiece movement that needs cleaning and restoration.

Fig 7.1 depicts the front and rear views of a movement that is considered to be typical of many that will be found to be in need of repair and restoration. We begin by taking a close look at the tools required for this task.

All the tools illustrated in Fig 7.2 were supplied by an English company from their catalogue which lists some 4,000 items ranging from clock pins to watchmaker's lathes and drilling machines.

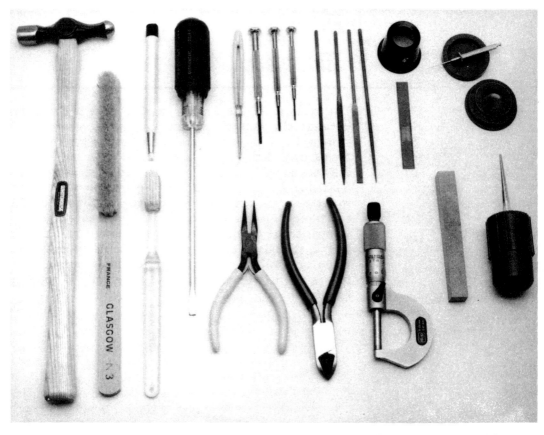

Screwdrivers

Several medium size screwdrivers are required for casework. For the movement, a set of small watchmakers screwdrivers with hexagonal handle ends and blade sizes ranging from about 0.8 mm to 3.8 mm would be ideal. The screwdriver is the basic tool of all repair work, and must be sharpened carefully if it is to do its job properly without damaging screw heads, or slipping and spoiling platework. One could be forgiven for thinking that the correct shape of blade is that labelled incorrect in Fig 7.3. The wedge shape in the same illustration is preferable. This form will not slip or disfigure the exposed edge of a screw head slot. A screwdriver blade must be kept sharp at all times; this is best achieved by placing it on an oilstone at the correct angle, and then moving it the full length of the stone without altering the angle. A scrubbing motion will only result in a blade with a rounded tip.

Fig 7.2 A suggested tool kit for cleaning and simple restoration work.   Top row (left to right): Hammer, buffing brush, fibreglass pencil, large screwdriver, tweezers, three small screwdrivers, four Swiss needle files, eyeglass with slotting file underneath, clock oiler and oiling cup with cover. Bottom row (right to left): Puffer, water-of-Ayre stone, micrometer, cutters, long nosed pliers, cleaning brush. All tools supplied by Southern Watch and Clock Supplies, Orpington, Kent, UK.

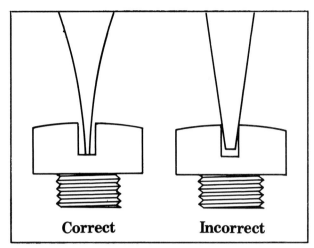

Fig 7.3 Correct (left) and incorrect shapes for a screwdriver blade. One with the correct profile will not slip and damage a screw head.

Correct   **Incorrect**

I have two sets of small screwdrivers, one marked with a white band of paint and kept for delicate work, and another set is used for stripping down movements and cases where the blades are often damaged by poorly slotted screw heads.

While using the screwdriver, a firm and even down pressure should be maintained to avoid damaged screw heads. Take great care with the humble screw; it can be difficult to replace if lost, and hard to refurbish successfully once the slot is damaged. Remember that a marble clock movement mounted in its case has only the backplate and a few visible screw heads as an indication to the outside world of its condition.

Files

A small set of Swiss needle files will be invaluable for most of the repairs encountered on these clocks. They should be of the best quality that can be afforded, and consist of the following types:- Flat, triangular, half round, square, and warding. A small point to note is that the flat file will probably have a safety edge; this should be of plain smooth metal that will not damage adjacent work. Often these edges will be found to be rough and should be rubbed with a small oil stone.

An unusual yet useful file, rarely seen outside the horological world, is the bull's foot file in Fig 7.4, which is used for filing bushes flush without damaging the surrounding plate.

Fig 7.4
Bull's foot file.

Finally, a slotting file is available for either cutting slots in new screw heads, or for repairing damaged slots in old screws. In the absence of the proper tool, an improvised file may be formed by taking a razor blade of the type with one reinforced metal edge, holding the sharp edge against a file and hitting the reinforced edge with a hammer.

Do not be tempted to use a file card to clean the files as the steel bristles will permanently damage them. The best way to clean files is to degrease them in petrol and then clean across the blade with a piece of thin sheet brass.

Pliers and Cutters

The best pliers to use in repair work are the long nosed type, preferably having smooth jaws; ideally the jaws should be brass-lined to protect polished surfaces. When buying pliers, hold them up to the light to check the blades. No light should be seen between the tips. (See Fig 7.5). A pair with an overall length of 140 mm and a larger 170 mm round nosed pair with serrated jaws for removing seized pins should be purchased.

Side cutters are used for cutting pins to size, and with care may be used to remove clock hands. Again an overall length of 140 mm is ideal.

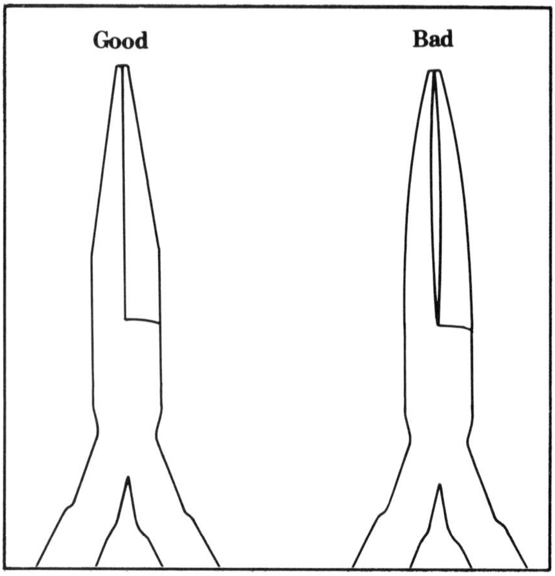

Fig. 7.5 Checking the quality of pliers. There should not be any light visible through a good pair of pliers.

Taps and Dies

With luck, the threads on screws and their relevant holes will inevitably be occasion to rethread a hole or make a screw, for example, making a threaded barrel hook, or re-threading the screw holes of the clickwork. A full tap and die set displayed in a polished wooden case is an expensive item to buy. Fortunately they may be purchased singly and a small selection will serve for most problems encountered. The principle sizes of tap and die required will be from about 1.0 millimetres to 3.0 millimetres.

The traditional tool for making clock screws is the screwplate, illustrated in Fig 7.6. It may still be purchased, with a set of taps for threading clock holes. The alternative and more modern method is to use taps and dies also shown in Fig 7.6. The bottoming tap is used for finishing blind holes, and the tapered tap for

110

conventional plated work and starting a thread in a blind hole. For normal work the tap holder holds the tool, although for small fine holes it is prone to eccentricity and a pin vice will provide more accurate support.

There is an element of confusion in the sizing of taps and dies. The type of thread normally encountered in new work is called 'metric coarse', this identifies the overall thread type. Each tap and die is stamped with a code. For example, a small size could have '4.5 m 0.75' stamped on it. There are variations: For example, one might be marked 'm 4.5 x 0.75'. The letter m denotes millimetres and the figures after or before the m, 4.5 in

Fig 7.6 Thread cutting tools, left to right: (1) Traditional screw plate, (2) Die holder, (3) Die, (4) Tap holder, (5) Tapered tap, (6) Bottoming tap.

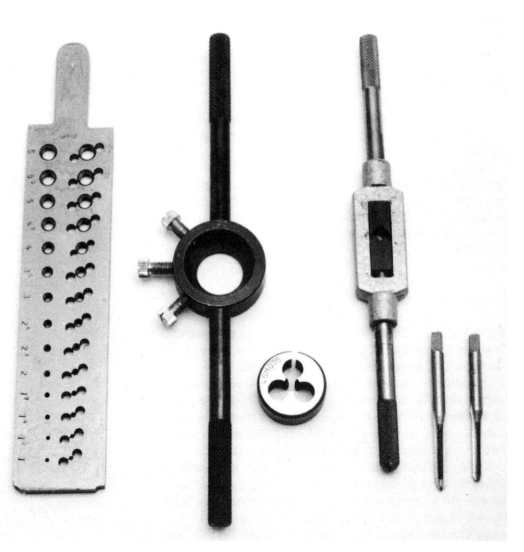

this case, give the diameter of the hole that will be threaded. The 0.75 is the number of threads per millimetre, and is the pitch of the thread. The distance between each crest on the thread is 0.75 millimetres, a pitch of 1.0 would give 1.0 millimetre between threads.

Tweezers

The requirements for tweezers are similar to those for pliers. Tweezers are normally used by watch repairers, but the delicate movements of the French clock require their use for the placing of small components, and also for reaching inside partially assembled movements to align pivots with their adjacent holes. When buying a pair, hold them up to the light to check that the tips come together and that the blades are laterally aligned.

Hammers

A hammer may appear to be a surprising choice of tool for repairing a clock mechanism, but it has a multitude of uses.

Hammers are sold by the weight of their heads and for work on clocks 100 grams will suffice. An engineers' ball-peen hammer will be most useful. The handle should be long; about 330 mm gives good control. The hammer face must always be kept clean and bright and an occasional burnish with emery cloth will help to avoid slippage.

Eyeglasses

Although clock repairers will not need to use an eyeglass or loupe for prolonged periods, considerable discomfort can be caused if the glass is not chosen with care. Sizes of the human eye socket vary from one individual to another, and for this reason an eyeglass should be tried for fit before purchase. My first eyeglass was bought without taking this simple precaution and was so large that it was impossible to hold it to the eye unaided.

Various focal lengths are available allowing a different distance between the object being focussed and the eye. An 83 mm glass will require the work to be held 17 mm closer to the eye than a 100 mm model. The magnification also varies with the focal length; an

83 mm lens magnifies by a factor of 3x against 2.5x of the 100 mm lens.

For French clocks, the 100 mm lens will be found to be the most comfortable, the shorter lengths being reserved mainly for watch repairing. Wearers of spectacles are, for once, better placed for a choice of eyeglass than those with unimpaired vision. A wide variety of hinged lenses may be found that clip neatly on to the frame of spectacles, allowing the glass to be quickly flipped in and out of place. Once again reference to a supplier's catalogue will reveal a large choice.

Brushes

An assortment of brushes is essential for cleaning and polishing. Polished wheels and plates will need to be carefully cleaned if the final quality of finish is to be up to standard.

Three brushes will satisfy most requirements. Two soft natural bristle toothbrushes (not synthetic) with long well-packed heads will do for cleaning while a larger soft brush (Glasgow No 3) will bring up the sheen on brasswork during the final buffing process. It is worth taking the trouble to buy new toothbrushes because the bristles will be firm, upright, and able to access some of the tiny crevices in pinions and wheels. One other item deserving mention is the fibre-glass pencil, a brush made of small glass fibres, held in a propelling pencil type of holder. It is useful for cleaning oil sinks, and rust spots from steelwork.

Clock Oilers and Oil Cups

Oiling, dealt with later in this chapter, is one of the most misunderstood aspects of repair work. When oiling the clock, a small drop of oil is picked up from an oil cup with an oiler and applied to the oil sinks that are hollowed into the back and front plates of the movement. Every time the oiler is used, it should be wiped on a clean piece of linen cloth. Some workers prefer to use the soft pithy wood that comes from the elderberry bush, which is dried, cut, and bundled as sticks, to be sold under the name of 'pithwood'. For cleaning, the sticks are cut into short lengths and packed end-on in a small tube about an inch high. The clock oiler is plunged in periodically to

113

clean it. Clock repairers often used an old French barrel as a container.

When oiling, some oil should be poured into a small glass or agate cup and covered immediately to avoid contamination from dust and other foreign matter. In an emergency, oil can be picked up with a small screwdriver direct from the bottle of oil, but this is not to be recommended because the whole bottle of oil can become contaminated and there is a genuine risk of knocking it over.

The Puffer

A puffer produces a small jet of air and, although not a necessity, is handy for blowing out fine pivot holes and removing brass dust from filed work.

The Vice

An engineer's vice in one form or another is a necessity. The type illustrated has steel jaws 90 mm wide. Most vices have sharp, serrated jaws that will damage soft

Fig 7.7 Traditional engineers' vice with 90mm jaws and, in front, home-made aluminium cheeks that fit over the jaws to avoid damage to parts held in the vice.

brass. In every case vice cheeks (also called 'clams') will be needed. They are easily fashioned from copper or soft aluminium sheet, as shown in the foreground of Fig 7.7.

I prefer the large vice with 90 mm jaws, although 75 mm would be perfectly adequate. The weight of a large one enables it to be used without being bolted to the bench. This portability is useful if, say, a well-lit desk top is used instead of the bench for fine work. With the jaws open, it also makes a temporary movement test stand and thirdly, the casting is strong enough to withstand use as an anvil for light work.

Woodworkers' vices are of little value to the horologist. They lack the fine screw thread and general precision of the engineers' model.

Micrometer

A micrometer is by no means an essential part of a clock repairer's toolkit, but is invaluable for measuring some of the tiny parts. Its use is essential for measuring the diameters of pins when restoring a Brocot pin-pallet escapement. Micrometers are quite expensive, but will last a lifetime. If buying a used model, close the jaws until they just touch, then check that the scale reads zero. If there is any substantial variation from zero, do not buy. Micrometer scales are in either Imperial or metric. Imperial measurements are still favoured by the model engineering community in Britain, but the metric system version is essential for parts for French clocks.

The Turns

Turns (Fig 7.8) may be difficulty to obtain, although second-hand versions may usually be found in one of the clock material dealers' shops in the Clerkenwell Road, London. They were used by clockmakers and watchmakers before the watchmakers' lathe. They are not essential items but often referred to in books.

With turns, the work is turned between dead centres to produce almost total accuracy, even when the item is turned end to end. They are still used by modern craftsmen for accurate work where the slight wobble of a lathe chuck, which has a bearing, would produce unacceptable errors. If turns are not available second hand, they are fairly easy to make using blocks of brass or mild steel.

115

Fig 7.8 Watchmakers' turns,
suitable for the delicate work.
The length of the bed is 160mm.
In front are specialised
tools for use in the turns.

Clock Cleaning Fluid

Fortunately few fluids are required for cleaning and
many can either be purchased cheaply or made up from
existing domestic products (Fig 7.9). With a few notable
exceptions a soapy ammonia solution is used to degrease
and clean most components of the clock movement.
Application of clock cleaning fluid is discussed later in
Chapter Nine under Cleaning. The liquid may be
purchased (e.g. Horolene) in concentrated form and
diluted with water. The resulting 'soup', as it is
sometimes called, may also be used for cleaning brass
bezels and various other brass case parts.

A cleaning liquid can be made by mixing an egg cup
full of cloudy ammonia with a litre of water and keeping
the prepared solution in a tightly lidded plastic container
with a large 'Poisonous' label attached. Add the ammonia
to the water first. In its undiluted state ammonia gives
off a noxious vapour.

116

Fig 7.9 Restoration materials, left to right: (1) Clock cleaning fluid, (2) Brass polish, (3) Clock oils, (5) Bundle of pegwood (in foreground).

Metal Cleaner

Brass plates and wheels are polished using a domestic brass cleaner. Brands such as Bluebell and Brasso are adequate. It is possible to buy a long term brass polish that will keep its shine longer and is particularly good for heavily soiled work and for the back plate, as that is visible from outside the case.

Blueing Salts

Blueing salts are not often encountered outside the horological trade. Whiteish in colour, the salts are supplied in a glass or plastic jar. In use, they are melted down in an old saucepan with a well-fitting lid and remain liquid when heated to extremely high temperatures. Bright and clean pieces of steel such as clock hands may be placed in the hot solution and held there until the correct shade of blue is achieved. A quick, professional finish is readily obtained by the less experienced worker.

Salts are relatively expensive to buy, but last a long time. Exercise great caution with both cold and hot

117

crystals. In their fluid state they are very hot and can cause serious burns, despite looking deceptively cool. Children should always be shut out of the room when the salts are hot. Any item being blued must be bone dry as a small amount of water will cause the hot salts to spit viciously. Above all, do not be tempted to stand them on a portable camping stove which has the burner on the can of gas.

Pegwood

Pegwood is an inexpensive hardwood sold in bundles. A stick, sharpened to a point, is used extensively for cleaning pivot holes, oil sinks and a wide range of other items, although matchsticks are preferable for the smallest holes.

Solder

Two types of solder are encountered in horological work: soft or lead solder, and hard silver solder. Soft solder melts at relatively low temperatures and is commonly applied with a soldering iron for electrical connections. Silver solder makes a stronger joint but requires the metal to be heated to red-heat. The best method of using either solder is to heat the item to be soldered with a small butane gas torch. The flame is adjustable at the turn of a knob from a thin pencil to a broad one.

Oils and Greases

Clock material catalogues list a bewildering variety of clock oils and greases. There often seems to be little difference between descriptions; for example, Microtime oils is 'an old favourite', while Windles has 'still the same proven quality'. Each repairer will have his own favourite, or gain one in due time. Oils must be purchased fresh, and stored away from daylight in a cool, dark place.

In some quarters it is recommended that a special lubricant, such as graphite oil, be used for mainsprings instead of the traditional drop or two of clock oil. The argument is that the power curves of springs treated with one of them are much more even than those of springs treated with pure oil. The uses of various lubricants recommended by one manufacturer, Moebius, are listed at the end of the chapter as a guide, although not necessarily to recommend these products against other oils. There is one golden rule to remember, use a light oil for small pivots, a heavier oil for larger pivots.

Fig 7.10
Removing one of the strap screws.

Removing the Movement from the Case

The first task facing the restorer is to remove the movement from the clock case. It is fortunate that cases are heavy and the weight helps to keep the case still when the movement is being taken out. Open the rear bezel cover, and, if the pendulum is still attached, reach inside and gently unhook it from the suspension spring. This spring is delicate and easily damaged by the slightest twist. If the pendulum is not visible, it may have dropped down inside the case. Inspect the pendulum under a good light with an eyeglass. The bob may have a maker's stamp on it that will help in the identification of an otherwise unmarked movement.

Where the top of the bob abuts the shaft there may be a small nick or file mark, made by a previous repairer to locate the exact position of the bob when it is being re-assembled. If not, make a small neat mark on the rear side of the shaft with a Swiss file.

Two long screws hold the movement in position, pulling it up against the rear bezel. The safest way to remove the movement from a very large case is to place the clock with its front glass flat on the workbench. Make sure the bench is covered by a cloth. Undo the two rear screws, and remove the rear bezel. The straps inside the case may then be held firmly while the case is gently turned upright, at the same time holding the front bezel

119

firmly against the case. Now pull the movement forward out of the case and place it gently, face downwards, on a bench protected by a cork mat or cloth.

With smaller timepieces, one of your hands will usually be able to span the case sideways, gripping the front and rear bezels at the same time. This leaves your other hand free to release the screws as shown in Fig 7.10. Remove the case to a safe place and settle down for a close inspection of the movement.

One of the small pleasures of repair work is looking with an eyeglass for the signatures or initials of previous repairers. A date mark is often seen written in a fine hand with a pointed instrument. Typically it will be written in the form 30/11/89, (where 89 is the year in question). Since French marble clocks were not around in 1789, it can be assumed that this date is 1889, although this is not a hard and fast rule. Clock repairers often mark their work to provide a reference date for their guarantee to a customer.

Dismantling the Movement

When dismantling the movement you will need the small screwdrivers, side cutters, long nosed pliers, and a cork mat to protect the movement. (A mat has not been shown in the accompanying illustrations because it interfered with photography.)

Step 1. Fig 7.11. The front bezel. The first item to be removed is the front bezel and glass. With a suitable screwdriver, carefully remove the small countersunk screws securing the bezel to the movement. Sometimes these are found damaged and with splayed slots. With care the slot can be deepened with a slotting file. This inevitably causes some cutting into the brass of the bezel; smooth it carefully at a later stage with a Swiss file. If re-slotting fails to remove the screw, it will have to be drilled out. Remove a screw from elsewhere on the

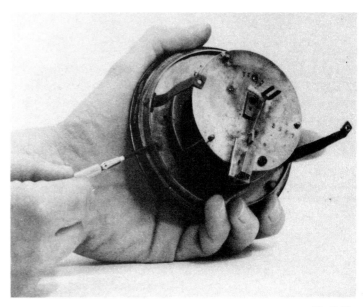

Fig 7.11
Removing the front bezel
retaining screws.

Fig 7.12
*Carefully lift the movement
clear of the front bezel.*

Fig 7.13
*Removing the hands, usually
using a small pair of cutters.
Fingers may also be used.*

bezel, and use this to measure the correct drill diameter.

Step 2. Fig 7.12. With the bezel screws removed, carefully lift the movement away from the bezel.

Step 3. Fig 7.13. Removing the hands. With a pair of long-nosed pliers, remove the pin holding the hands in position, remembering that it is a tapered pin and will only come out one way. The hands should now be removed one at a time. Sometimes the fingers will do this, but more often some form of implement will be required. With care small side cutters can be used; gently slide them under the base of the hand; then ease it away from the dial.

If the hands are particularly hard to remove, take a screwdriver with a 5 mm blade and place it at the base of the hands and between them. A gentle twist should remove the minute hand and the hour will almost certainly follow.

From now on, whenever the movement is put down, place it face downwards on a small cloth-covered dish or ashtray to protect the arbor that the hands were mounted on.

Step 4. Removing the dial plate. With a pair of long-nosed pliers remove the pins securing the dial plate to the movement. Lift the movement clear, and place it to one side for a while. With the eyeglass, take a close look at the front plate, where most repairers will have left their signatures.

121

Fig 7.14
Letting down the mainspring.

Step 5. Fig 7.14. Letting down the mainspring. If you have made the mainspring winder shown later in Fig 7.22, the handle from the drill can be fashioned into a let-down tool simply by making a saw cut in one end, into which the handle of the key may be placed.

Most movements will have stopped with the mainspring fully wound by unwitting owners who hoped that a good wind would cure all problems. Unfortunately overwinding causes some of the worst damage encountered in an otherwise good movement.

Hold the movement in one hand in such a position that your thumb can activate the clickwork. Take a well-fitting key, insert it into the let-down tool, if available, and move the winding square in a clockwise direction as if trying to wind the clock up. With pressure maintained by the key, release the click and allow the spring to unwind slowly. Do not underestimate the power of the spring. If it is accidentally released, you will at best suffer grazed knuckles, but worse, cause damage to the movement. If a key alone is being used, proceed in half turns. However, with the letdown tool, gentle pressure can be maintained on the wooden handle while the spring runs itself down.

Step 6. Preparing a sketch. A sketch should now be made of the movement, indicating the relative positions of the train wheels and the exterior components. Ideally a repairer should know the exact function of each component. With a timepiece the situation is fairly straightforward, but a striking clock is much more complex and it is therefore good practice to cultivate the habit of making notes. Bear in mind that many days or weeks may elapse before the movement is re-assembled. Photographs are of some value, but not so good for indicating the relative positions of the train components.

The Initial Inspection

This is an ideal opportunity to inspect the movement for obvious signs of wear and damage before dismantling it completely. On pendulum movements it is common to find the suspension spring broken or twisted. The slightest kink will disrupt the pendulum's true motion, affecting timekeeping and overall performance. Suspensions are available now with either brass or plastic ends. In many ways the plastic type are preferable, as the brass ones can be of doubtful quality.

Continuing with the inspection of the movement, the following points should be noted: Poor workmanship in an earlier repair, wear, and missing components. Poor workmanship will probably be the biggest single headache. Most movements were made to a high standard, but subsequent repairers may have induced problems. Most examples of slipshod repair work will have been performed to save time, rather than through lack of craft skills and one should not be too critical. Repairers are often expected merely to get a clock going because some owners will not spend a reasonable sum of money for having the correct work performed.

Fig 7.15 and 7.16 give an idea of the places on a movement that should be checked for wear or damage.

Fig 7.15
Where to check for wear on the front plate:
(a) Escapement arbor pivot hole,
(b) Loose click retaining screw with damaged thread,
(c) Damaged or missing click wheel teeth.

123

The taper pins that hold the plates and bezels together are a good indication of previous craftsmanship. Ideally the pins will always be straight and in good condition, although old pins will inevitably have been used more than once. The ubiquitous dressmaker's pin, used instead of a taper pin, will have often found a resting place usually holding the hands in position or retaining the striking hammer. Check carefully for any odd-looking screws. These could have the wrong thread for the hole and may not be replaceable during reassembly.

Wear is not often a serious problem, although a cursory glance over the wheels and pinions may show evidence of it. The click wheel on the front plate should be checked for broken or missing teeth (c in Fig 7.15). Still at the front plate, look at the winding square, which will sometimes have rounded edges caused by an ill-fitting winding key. Fortunately the defect is easily rectified. If a Brocot or Vallet pendulum regulator is fitted, check that turning the shaft causes the suspension spring or Brocot cheeks to move. Do not force the shaft; it may have become jammed.

It is not uncommom to find click wheels and other external parts missing. This is a risk that is taken when buying an old clock and is a good reason for buying from a shop rather than an auction house, because at least you will have some recourse if a simple guarantee was obtained.

Checking for Wear

An eyeglass and a desk lamp is used to check all pivot holes for excessive wear. Hold each arbor in turn and gently rock it up and down while looking for wear in the plate hole. In severe cases, the hole will have become oval in shape (Fig 7.17). It is difficult to describe exactly what constitutes unacceptable wear; only experience and commonsense can help. The barrel holes are checked in a similar way, by holding the barrel between two fingers and moving the winding arbor from side to side.

If excessive wear is found, mark the relevant holes with a circle drawn around them, using a black felt tip pen. If this simple precaution is not observed, there is a danger of repairing the wrong hole at a later date.

124

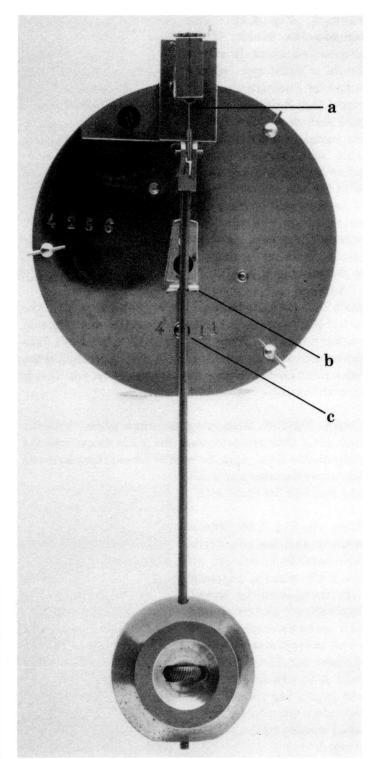

Fig 7.16 Check for wear on the back plate. (a) Damaged suspension spring, (b) Loose pendulum crutch, (c) Wear in a winding arbor pivot hole. (It is just behind the pendulum rod in the picture). Check both plates for this.

a

b

c

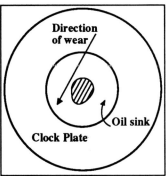

Fig 7.17 A pivot hole that has become oval in shape due to excessive wear.

Direction of wear

Oil sink

Clock Plate

Step 7. Fig 7.18. The suspension block. If the suspension block is of the Brocot or Vallet type, remove it now by loosening the large screw that holds it and the back cock in position. Lift off the complete block assembly while still holding the pallet arbor with the other hand. With the suspension block out of the way, gently remove the pallet arbor and attached crutch. Take care as these pivots are among the smallest in the movement.

Remove all remaining parts from both the back and front plates except for the cocks, then remove the pins holding the rear plate in position, remembering to place the movement face downwards on a deep ashtray with a cloth protecting the front plate. Do not remove the cannon pinion yet. This is done in Step 10.

Fig 7.18 Removing the suspension block.

Step 9. Fig 7.19. Removing the back plate. With the back plate pins removed, ease the plate away from the pillars of the front plate. Be careful when lifting it free as there may be some pivots left gummed into the back plate.

Fig 7.19 Gently easing the back plate away from the movement.

Step 10. Fig 7.20. Train wheels and barrel. It is now possible to remove all the train wheels, together with the mainspring barrel. The cannon pinion should pull away from the centre wheel using just your fingers. If, however, it is jammed, hold it firmly in one hand and gently tap the arbor of the centre wheel once with a small wooden mallet or block of wood.

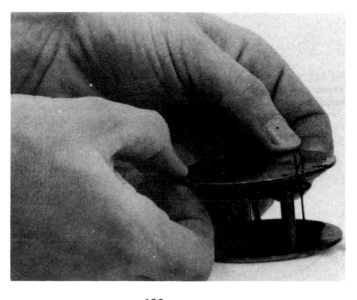

Mainspring and Barrel

The clock mainspring is contained within a brass barrel and may either be removed by hand or preferably with a mainspring winder (Fig 7.21). Unfortunately, even the simplest winder is expensive, equal at the time of writing to the price paid for the clock illustrated. Removing springs by hand is not to be recommended. They usually become distorted. Obviously a low cost mainspring winder is required that does not require extensive machining or metal work skills in its manufacture. In fact a cheap conventional hand drill can be converted into a winder as in Fig 7.22.

A punched mark can sometimes be seen adjacent to the opening in the barrel cover. If it is not there, make a neat mark to enable correct re-orientation of the cover

Fig 7.20 Removing the mainspring barrel.

Fig 7.21 A traditional mainspring winder, made by Colin Walton Clocks, England.

during re-assembly. Insert the largest screwdriver possible into the cover slot, and prise it off.

The mainspring arbor may now be unhooked and removed from the barrel. Examine the pivots on the arbor with an eyeglass, making a note of their condition for possible repair. A close look at the edges of the winding square will often reveal damage caused by an ill-fitting key. Rounded squares are easily repaired (Fig 7.23), by holding one face flat on a steel stake, or large hammer head. Hammer each flat in turn until the squareness is restored; then clean up with a file.

Fig 7.22 Improvised winder made from an inexpensive hand drill and only suitable for small springs.

Removing a Spring by Hand

If a mainspring has to be removed by hand (despite being unwise), hold the barrel in one hand and, with the smooth jawed long-nose pliers, pull out the middle of the spring until one coil is free and starts to coil over its neighbour. Grip the protruding piece of spring and ease it out from the barrel. Great care must be exercised as the spring can very easily become distorted.

*Fig. 7.23 Flattening the squares
of a winding arbor.*

Using a Spring Winder

To remove the spring with the help of a winder, take a piece of cloth and hold the teeth of the barrel with it. Offer up the barrel to the arbor of the winder and adjust the position of both until the spring eye has hooked. Grip the barrel firmly and rotate the wheel of the winder until the spring is clear· of the barrel walls. Now carefully insert the tips of the smooth jawed pliers into the barrel, and grip the outer end of the spring. Gently rock the barrel until it is unhooked from the spring eye, and then remove it. Now take the cloth again and hold the coiled spring. Release the pliers and slowly allow the spring to uncoil inside the cloth. It would be wise to wear safety goggles; it is surprising how much harm an accidentally released spring can cause.

Cleaning and Inspecting the Spring

With the spring removed, take a clean cloth and place it between the jaws of the long nosed pliers. Lightly grip the spring and wipe it from one end to the other. The centre coil can be cleaned with the cloth wrapped around a stick of pegwood. Do not use any form of solvent on the surface of the spring.

The eyes in the spring should now be inspected and checked for signs of wear or cracking. Place the spring edge down on a piece of white paper; its shape should be like that in Fig 4.1, not that in Fig 7.24. If the coiled spring is barrel bound (too tight), replace it. If the spring requires a new hole, follow the procedure shown in Fig 7.25 to Fig 7.28.

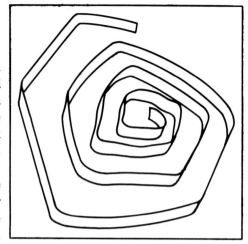

Fig 7.24 A barrel bound spring (exaggerated); a good spring will be a smoothly shaped spiral.

Checking for an Oversize Spring

Ideally a mainspring should occupy about one-third of the barrel. Fig 7.29 shows a simple tool made of stiff card that may be used for checking this dimension. If the spring is too large it will undoubtedly create problems.

The mainspring barrel and cover should be closely examined for evidence of the spring edges rubbing against these surfaces. It is possible that a slightly oversize spring was fitted during an earlier repair and it is unwise to assume that the current spring is of the correct size. Check both the spring height and the barrel's internal height and diameter. When checking the internal depth of the barrel, remember to allow for the cover recess.

There is a tendency to attempt to overcome problems by fitting a slightly larger and therefore more powerful spring. This should be avoided at all costs; it is best to take the trouble to check for the cause of the problem.

The Barrel

While the spring is removed, the hook inside the barrel can be inspected. This should protrude from the wall of the barrel as little as possible without losing its hooking

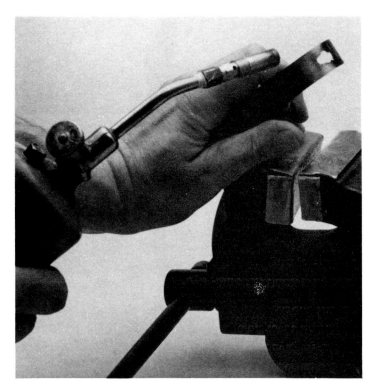

Fig. 7.25 Renewing the eye hole on a mainspring: Annealing the metal about 50mm from the outer end before making a new hole in a mainspring.

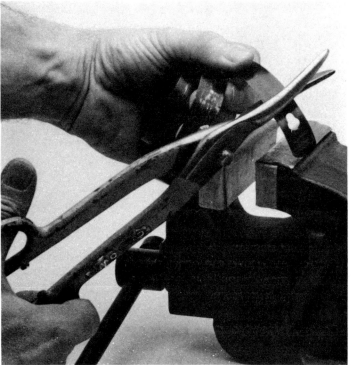

Fig. 7.26 Cutting off the damaged section.

131

Fig. 7.27 Punching a hole
in the end of a mainspring
using either a proprietary spring
hole maker or a punch
and pieces of hardened steel.

Fig. 7.28 File both surfaces
smooth and make the hole
slightly oval.

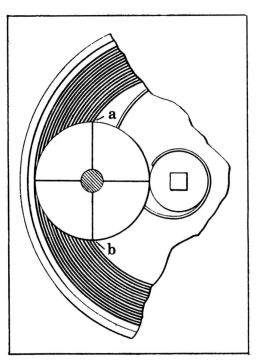

Fig. 7.29 A simple device made out of a disc of cardboard that will help to see if a new spring is of the correct size.
The spring should intersect at points (a) and (b).

effect. If it takes up too much room, it reduces the area into which the spring can uncoil, thereby limiting the power output. Hooks are often found to be loose through overwinding, or for the same reason to have distorted the barrel wall. Fortunately this situation is easily remedied. Fig 7.30 to Fig 7.33 show the method of replacing the hook and repairing any damage to the barrel.

Another result of overwinding is that the barrel teeth can become distorted out of position by the sheer force of the key being wound against a seized movement.

Fig 7.34 illustrates the method of easing the teeth back into position. Unfortunately the teeth may still break off when this is done. If several teeth are broken, they will have to be cut out and replaced with a keyed piece of brass. If the reader has access to a lathe and dividing head, it may be better to replace the barrel wheel completely. Alternatively a suitably qualified clock restorer should be consulted; the cost of his work may be well worth the time and effort involved. If this course of action is decided upon, present the restorer with the complete barrel assembly minus the spring. The correct approach will invariably solicit a great deal of help, but do not expect him to get excited about French marble clocks - they are still frowned upon in some quarters.

Occasionally advertisements will be seen in clock magazines for the supply of barrels and cutting of wheels.

Repairing Pivots

Fortunately the smalller pivots are hard and do not wear excessively, although they are prone to damage by snapping, or bending. It is sometimes possible to straighten a bent pivot using pliers or tweezers, but first the metal will need to be softened (annealed) by heating it to a cherry red colour and then cooling slowly. The pivot is best wrapped in iron wire to help slow down the cooling process. After annealing, straighten it with pliers and check if it runs true by spinning it in a pin vice. Continue straightening until success is achieved. The metal is rehardened by repeating the heating process

133

Fig. 7.30 First drill out the old hook when replacing a barrel hook.

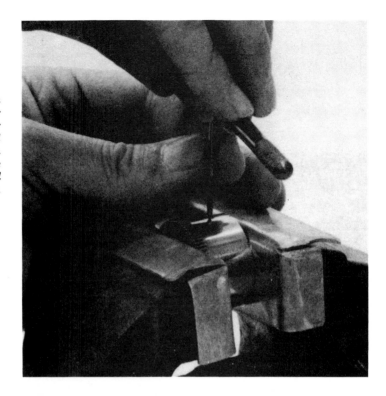

Fig. 7.31 Tap a thread into a new hole for a barrel hook, or close the old one with a hammer and stake. Insert a small countersunk headed bolt from the inside. Before inserting the bolt, ensure that most of its head has been filed away.

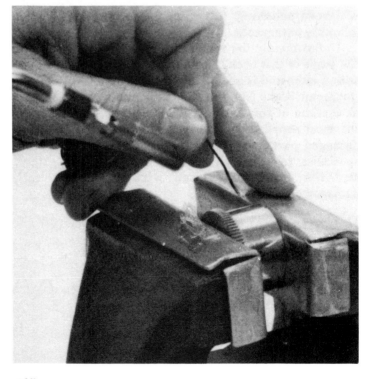

Fig. 7.32 Carefully solder the bolt to the brass barrel.

Fig. 7.33 Remove the exposed thread with a file.

and then quenching rapidly in oil. After hardening, it will need polishing, following the instructions in the following paragraphs.

Unfortunately the repair of a broken pivot is beyond the scope of this book. It is a fairly straightforward task with a lathe and is adequately described in the repair books found in the bibliography at the end of this book. Without a lathe, the most sensible solution is to send the damaged part away to be repaired by a specialist, together with the clock plates and the wheel associated with the damaged pivot.

Polishing Pivots

Clocks with tiny pivots and small mainsprings require every scrap of power available if they are to function correctly. Most power is lost by friction between the pivots and their bearings. This can be minimised by polishing the pivots with a special tool, the burnisher, until a fine

Fig 7.34 Easing damaged barrel teeth back into position. This should be undertaken with great care.

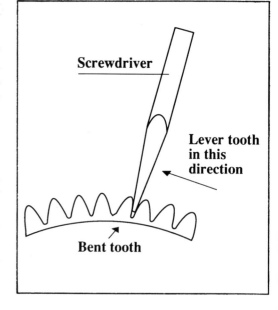

sheen is achieved.

Many books will suggest that it is possible to perform this work by using a pin vice to hold and spin the pivot, using a block of wood as a rest; this is not good practice. The work should be turned between dead centres in a watchmaker's lathe, or in turns. I use a small pair of clockmaker's turns.

In the absence of a lathe or turns, a small compromise will have to be made. Pivots are not usually badly pitted and will benefit from being polished with a broad, flat piece of pegwood dipped into a liquid metal polish such as Brasso. Rotating the pivot by a pin vice, and resting it upon a wooden block should ensure that it is given a slight polish without damaging the truth of its bearing surface.

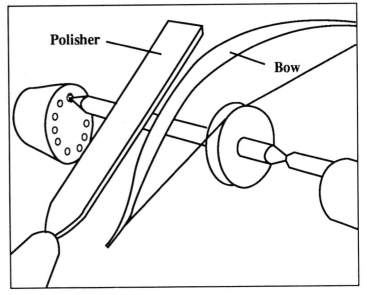

Fig 7.35 Set up for polishing an arbor in the turns.

Complete accuracy, however, is only possible with turns or a similar tool. The turns illustrated in Fig 7.8 are driven by a bow and line that is moved to and fro to rotate the arbor. Fig 7.35 shows the turns set up for use with a polisher made from a file that has had its abrasive surface ground off. A fine cutting surface is formed by rubbing an emery block across the smooth file face. Before attempting this work, consult one of the technical books in the Bibliography, such as those by de Carle, Gazeley, or Daniels.

After the pivots are repaired and polished, attention should be turned to repairing the holes in the plates in which they run.

137

Re-bushing Pivot Holes

Re-bushing pivot holes is possibly a task that an inexperienced reader may find daunting. However, it takes longer to read about than to do the job! It is important not to undertake any re-bushing work until the pivots have been repaired and polished.

It is possible merely to clean and oil a movement with worn pivots, and it will probably run for several years, although this is merely storing up worse problems for the future. Wear in the pivot holes is created by dirt and oil forming an abrasive paste that gradually wears away the brass plate and to some extent the steel pivots.

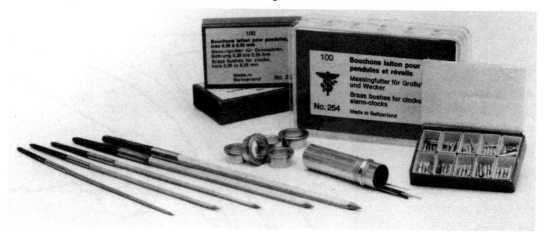

Tools and Method

There are two ways of re-bushing French clock movements. The least expensive involves the removal of brass from the clock plate with a hand tool, the broach, and the insertion of a small brass bush into the hole. Broaches are shown in Fig 7.36. A hole bored in the bush matches the size of the pivot. The second method involves the use of a relatively expensive device, the clock bushing tool, Fig 7.37. The use of this tool is shown in Fig 7.38 to 7.41.

Fig. 7.36 Tools and materials for rebushing holes. The small boxes contain little bushes - bouchons. In the middle are larger bushes, used for barrels. The two sets of needle-like tools, one set in a container tube, are broaches.

Using a Broach

A broach is a hand tool consisting of a steel rod with five flutes along its length. The steel is slightly tapered and the cutting action is provided by the sharp edges of the

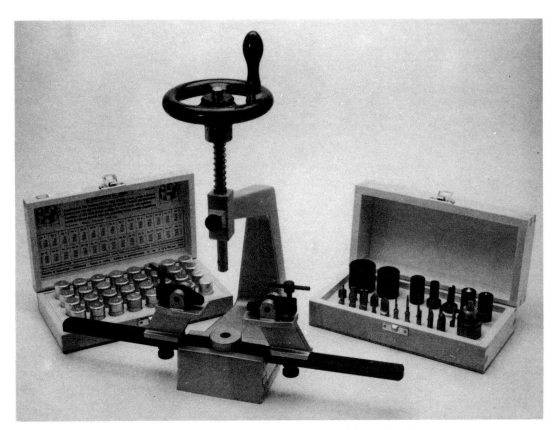

Fig 7.37 The Bergeon bushing tool. The box on the left contains an assortment of bushes. Tools supplied by Southern Watch and Clock Supplies, Orpington, UK.

flutes. The sizes of broaches refer to the diameters of the holes they cut. For French clocks, a selection from 0.7 mm to 6.0 mm plus a selection of very small, fine broaches called 'pivot files' will suffice. The main selection of broaches are clock cutting; there is another type, clock smoothing, for polishing and hardening newly cut holes.

An examination of the clock plate may reveal holes that require re-bushing other than those already marked. It is helpful if they are circled with a soft pencil or marker pen. With the profusion of holes it would be easy to find yourself broaching out the wrong one.

Fig 7.42 shows a pivot hole in a plate that has been 'punched up'. This vicious term is reserved for a technique that is not reommended. It involves the use of a punch, often a centre punch or a round headed one. To correct wear in the hole, the surrounding metal is squeezed towards the centre using the punch. The result is a bad bodge. Fortunately it is not often encountered in French movements, being seen more often in the plates of longcase clocks. Unfortunately punching up is usually

139

Fig. 7.38 Using the Bergeon bushing tool:
Centering the worn hole to be bushed.

140

Fig 7.39 A reamer is placed in position and the tool handle is rotated to cut the hole.

Fig 7.40 A bush is inserted in the hole and pressed into position.

Fig 7.41 Surplus material is cut away with a rose cutter.

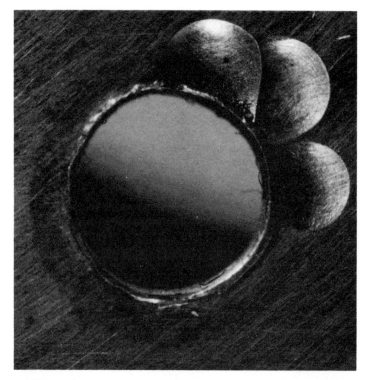

Fig 7.42 A barrel hole in a clock plate that has been punched-up to take up wear on one edge instead of being rebushed.

performed on the inside of a plate and is not normally visible when buying a clock.

One other bodging method, only slightly less frowned upon, is that of closing holes. It is performed by a special hollow punch that spreads the plate metal in from all edges of a hole.

The correct method of effecting this relatively simple repair is by bushing.

Fig 7.43 shows two types of bushes. The small ones are called 'bouchons' and some boxes are marked 'Bouchons laiton, avec manche' (Brass bushes with shaft). They are particularly useful for the smaller holes and make re-bushing an easily accomplished task. Referring to Fig 7.43 again, the larger part of the bouchon is the shaft, which is held with a pin vice. The other end is the bush itself; this is pushed into a pivot hole enlarged with a broach.

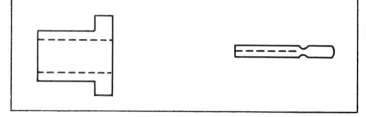

Fig 7.43 On the left is a bush and on the right a bouchon. The stub on the right hand end of the bouchon allows it to be gripped during insertion.

142

Re-bushing Small Pivot Holes

Fig 7.44 to Fig 7.46 illustrate the method. Having found a pivot hole that is in need of re-bushing, find a bouchon that has a hole slightly smaller than the diameter of the

Fig 7.44 Rebushing a hole with simple tools. Firstly, a broach is used carefully to open up the plate hole.

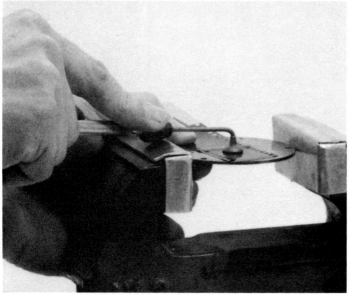

pivot. A broach is now placed in a pin vice. Insert the clock plate in an engineer's vice ensuring that the sharp jaws of the vice are covered with cheeks. A large drop of oil applied half way up the broach will slowly run into the pivot hole and make a considerable difference to the cutting action of the broach. Slowly open the hole, starting from the inside side of the plate, and keep checking its size with the bouchon. When the bouchon just starts to enter the hole, push it in as far as it will go. Ideally it will project slightly beyond the oil sink on the other side of the plate. Break off the shaft with a gentle twist of the pin vice and then tap the stump once with a punch.

Put the clock plate back in the vice with the inside facing upwards. The bull's foot file may now be used to file off the remaining stump of the bush. A piece of photographic film with a small hole cut in it, and used in the manner shown in Fig 7.46 will help prevent the scratching of the surrounding plate.

With the bush reduced to being almost flush with the plate, turn the plate over and inspect the oil sink. The bouchon should protrude beyond the pivot hole, and can be reduced either with an oil sink cutter or a suitably small twist drill held in a pin vice.

The next stage is to open the pivot hole from both sides until the pivot fits perfectly. A water-of-Ayr stone is finally used to hone the plate back to its original smooth surface. The stone should be kept wet either with running water from a tap, or by constantly dipping it into a container of water to which a drop of washing-up liquid has been added; this will help disperse the grease and oil that builds up on the plate.

The action of the pivot and its wheel should now be checked by replacing it between the clock plates together with its meshing wheel. Both should run freely, with the teeth meshing at the correct depth.

Chapter IX deals with the cleaning and rebuilding of a timepiece movement together with repairs to the pendulum and escapement. When this work is complete, refer to the lubrication chart on the next page.

Lubrication chart on next page

Lubrication Chart

This chart is designed to give a guide to lubricating the French movement. Although Moebius products are quoted here, there are other manufacturers who supply quality lubricants.

Moebius 8000 (or a pocket watch oil)
> Pallet arbor
> Escape wheel
> Escape wheel teeth
> Third wheel
> Suspension block thread
> Brocot pallet arbor thread
> Pendulum crutch cheeks
> Motion work

Moebius 8141 (or an oil designed for large chronometers)
> Intermediate wheel
> Centre wheel
> Barrel pivots
> Barrel cover hole
> Clickwork

Moebius 8200 (a semi-liquid mainspring lubricant)
> Mainspring
> Barrel walls

VIII

Restoring a Striking Movement

A striking movement strikes the hours and half-hours, normally on one bell or gong. A chiming movement sounds the quarter hours along with the hours and half-hours on more than one bell. The restoration of a striking movement from a French marble clock will provide the amateur with a demanding yet potentially satisfying task. I recommend rereading the relevant sections of Chapter Six with a movement in front of you, unless you have worked on this type of clock before.

It would be advisable for a novice to work on several timepieces before tackling a striking clock. A methodical and careful approach to the job will usually overcome most problems, especially if difficulties are encountered with temperamental striking.

Two main striking systems are employed: rack and locking plate. The rack striking movement is the more complex of the two, as already explained, and the number of hours struck will always tally with the position of the hands even if the striking train runs down and is rewound at any time later. The locking plate movement is slightly easier to work on, but if the striking train stops while the clock is still going and is rewound later, it will have to be manually advanced to the correct striking sequence that corresponds to the position of the hands.

Removing and Inspecting the Movement

Removing the movement follows the description in Chapter Seven, with the slight difference that we have to take into account the presence of the bell or gong. Take great care with gongs; they may appear to be fairly robust, but are prone to metal fatigue at the point where they join the main mounting block. Bells cause little trouble. They are simply removed from the back of the movement before being lifted away from the case. There is, incidentally, no need to unscrew a gong from the base board on which it is mounted.

To check rack striking (Fig 8.1 to 8.3), with the movement removed from the case, examine it with an

147

eyeglass, taking particular note of worn pivot holes, especially in the bell end of the strike train. Check to see how tight the hammer is on its arbor. There may be a lump of solder holding the hammer to its boss. On the front of the movement look for looseness of the striking components and make a note of any missing parts.

Fig 8.1 Front view of a rack striking movement.

With the locking plate movement (Fig 8.4 to 8.6), check the mounting of the hammer on its arbor. Look for wear in the pivot hole of the locking plate knife edge arbor, as well as the strike release lever bush. Hold the strike release lever while moving the knife edge to and fro because there may be some unwanted movement here. It is important to check the rear pivot hole of the locking plate wheel arbor; any wear there could give disturbing problems with the strike sequences.

Checking the Striking Work

I find it best to test the operation of the striking train before splitting the plates, although little can be achieved with a jammed train other than trying to establish the cause of the problem. With rack

148

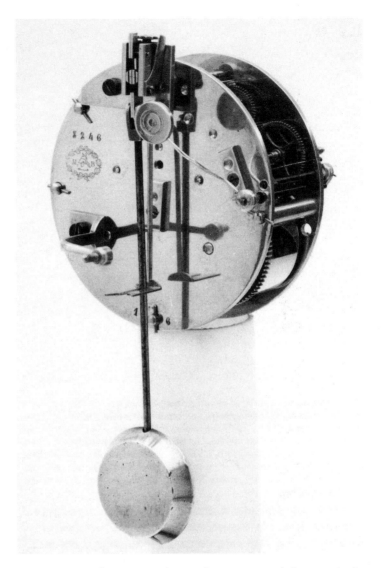

Fig 8.2 Rear view of a rack striking movement.

movements, the outer pin on the cannon pinion controls the initiation of the hour striking, while the inner pin looks after half-hours. The selection of half-hour or full hour striking is determined by the locking plate itself.

Try moving the minute hand round carefully and ensure that the pins on the cannon pinion are lifting their relevant levers and that the levers are dropping. The motion work will sometimes cause problems, especially if the hour wheel has moved too far forward on its mounting.

If all is well, check that the train strikes the hours and half hours correctly. Both types of movement are

149

Fig 8.3 Component parts of a rack striking movement. (1) Back plate, (2) Front plate, (3) Going barrel, (4) Strike barrel, (5) Escapement and pendulum crutch, (6) Pendulum suspension block, (7) Escape wheel, (8) Intermediate wheel, (9) Third wheel, (10) Centre wheel, (11) Warning wheel, (12) Gathering pallet wheel, (13) Pin wheel, (14) Intermediate wheel, (15) Hour wheel and hour snail, (16) Minute wheel, (17) Cannon pinion, (18) Minute wheel cock, (19) Fly, (20) Rack hook, (21) Strike stop arm and arbor, (22) Spring for [21]. (23) Lifting piece and warning lever, (24) Rack and rack tail, (25) Bell stand, (26) Hammer, (27) Hammer arbor, (28) Spring for [27], (29) Ratchet wheels, (30) Ratchet wheel cocks, (31) Clicks, (32) Click springs.

prone to the problem of striking only once, regardless of whether there is supposed to be a half or full hour sequence. This is usually caused by mis-alignment of the various arms that are activated by the lifting pieces. The most likely culprit on a rack movement is the rack hook; on a locking plate movement, the knife edge lifting lever may often be at fault.

This preliminary investigation usually proves to be time well spent; it is easy to rebuild a cleaned movement only to find that it did not function correctly in the first place. Always look for evidence of sloppy repair work. When rebuilding a striking movement, it is infuriating to find the newly-cleaned brass covered with fingerprints after trying to set up a temperamental strike train. Take time to make the preliminary checks before starting to strip the movement.

*Fig 8.4 Front view of a locking
plate movement.*

*Fig 8.5 Rear view of a locking
plate movement.*

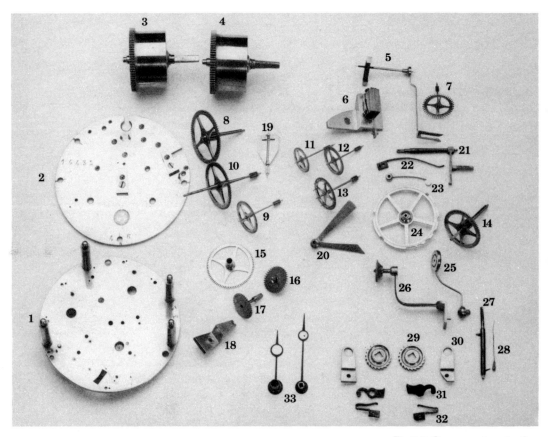

Take Notes Before Dismantling

Fig 8.6 Component parts of a locking plate movement: (1) Front plate, (2) Back plate, (3) Going barrel, (4) Strike barrel, (5) Escapement and pendulum crutch, (6) Pendulum suspension block, (7) Escape wheel, (8) Intermediate wheel, (9) Third wheel, (10) Centre wheel, (11) Warning wheel, (12) Strike stop wheel, (13) Pin wheel, (14) Locking plate wheel, (15) Hour wheel, (16) Minute wheel, (17) Cannon pinion, (18) Minute wheel cock, (19) Fly, (20) Lifting piece and warning lever, (21) Locking plate knife edge, (22) Knife edge lever spring, (23) Knife edge lifting lever, (24) Locking plate, (25) Hammer, (26) Bell stand, (27) Hammer arbor, (28) Spring for [27], (29) Ratchet wheels, (30) Ratchet wheel cocks, (31) Clicks, (32) Click springs, (33) Hands.

Before starting work, it would be wise to take note of a few points; failure to do so may result in problems when rebuilding the movement.

The position of the locking plate should be noted with respect to its arbor. The plate is squared on to this arbor and is designed to work in one position only. Sometimes there will be a mark cut into one of the squares on the arbor which corresponds with a punch mark on the back of the locking plate. If you are unable to locate these, make a neat cut with a file on one of the arbor faces and then another mark with a small punch on the rear of the locking plate. I omitted to take this fairly obvious precaution on the first locking plate movement that I restored and spent hours trying to sort out a strange striking problem.

With rack movements, make a careful note of the position of the gathering pallet on its arbor.

For all movements, make a sketch of the relative positions of the strike train components. This will prove invaluable when rebuilding. Make particular note of the positions of the wheels and the positions of the various pins that engage with them.

Some manufacturers will have helped in this direction by making punch marks on wheels indicating where they should mate during normal operation. Check carefully to see if this is the case with your movement.

Dismantling the Movement

Coping with seized parts is often a problem with a neglected movement. The techniques employed for locking plate movements are almost identical to those described in the last chapter. However, more components are squared on to arbors and may have become rusted or frozen into position. A little releasing fluid, maybe some heat, and a lot of patience will help in removing them. Take great care not to bend any of the brass arms; they are very fine and easily break. Always try to use two levers opposite each other under the base of such an arm when levering it off. Lacking special levers, use the blades of two small screwdrivers. Above all do not be heavy handed.

Marking Components

When removing parts, especially those that are similar, make sure that they are clearly marked. There are, for example, two barrels, one for the mainspring and the other for the striking train spring. Often they will be marked with an S or G denoting strike or going. 'Going' is, of course, the name given to the timekeeping train. Similarly the click wheels and their associated springs will need to be marked. The springs must be identified when they are removed from their barrels as their characteristics are often quite different, although this is not always visible to the naked eye.

Repairs and Replacements

Wear is an important factor to remember when looking at the strike train. When assessing wear on a timepiece

153

movement it is usually possible to make a subjective decision as to whether repair work is really necessary. In a striking movement, however, wear will induce some strange problems. For example, take the arbor on to which the hammer arm is squared. If the plate hole that it runs in is worn, the hammer will tend to bounce when it strikes the gong or bell. This may not always be noticeable to the untrained ear, but one quickly becomes attuned to the proper sound. With a bell, a tizz-tizz note is heard instead of a clear and sweet ting-ting. Gongs can sound even worse, sometimes giving the impression that they are being hit with the blade of a knife.

Occasionally detent springs will be missing or broken, but do not worry, because material dealers stock a surprisingly wide range of springs.

Missing Strike Work

It is possible to make a replacement locking wheel if one has mysteriously disappeared. It is an exacting task that will give an insight into the amount of work that goes into the making of a simple movement. Similarly the various levers used in both rack and locking plate movements can be built if a little care and patience is used. Stiff card or thin metal sheet will be useful to make a template to check the approximate geometry of the component.

Finally, remember that there are secondhand movement suppliers who regularly advertise in the trade journals; they may be able to help with a missing part or even sell you a complete movement. Above all, do not be in too much of a hurry; it is amazing what will turn up just as you have given up all hope of ever finding an elusive part.

Cleaning the Movement

Cleaning methods are identical to those described in Chapter Nine. With double the number of components, there is a danger of skimping the detailed cleaning of oil sinks and pivot holes. Take time to do this job properly, using clean, sharp pegwood to clean out the larger holes and matchsticks for the fine ones. Make a point of cleaning the oil sinks either with pegwood covered with cloth or with a fibreglass pencil.

Fig 8.7 Rear view of a rack striking movement on which the brass back plate has been replaced with one made of clear Perspex: (a) Strike barrel and main wheel containing the driving spring, (b) Intermediate wheel, (c) Hammer arbor with its squared end, (d) Pin wheel, (e) Strike stop arm, (f) Gathering pallet wheel, (g) Warning wheel, (h) Fly.

Rebuilding the Movement

The task of rebuilding the movement demands a lot of care and patience. Someone who has never worked on one before might be wise to practice by reassembling the striking train only. This exercise will remove any additional problems that may occur with the going train.

The following notes on the reassembly of a rack striking movement relate to Fig 8.7.

155

1. Replace the centre wheel first, placing a smudge of oil on the front pivot before assembly. Similarly, oil the pivots of the barrel winding arbors before replacing the barrels.

2. Now mount the parts in this order: Gathering pallet wheel f (Fig 8.7); intermediate wheel b; warning wheel g; fly h; and pin wheel d.

3. Place a tiny amount of oil on the hammer arbor pivot c in Fig 8.7 and replace it with its pin resting between any two pins of the pin wheel. Follow along with the strike stop arm arbor e. The strike train should now be adjusted so that the pin on the strike stop wheel is resting on the upper edge of the strike stop arm. The pin on the warning wheel will need to be directly under the fly, i.e. the pin should be as near to the fly arbor as possible. This will give the warning wheel its required half-turn before being caught by the strike release lever.

4. The back plate, without its cocks, may now be carefully replaced. Start with the pivots nearest to the barrel and then work out towards the others. Eventually the plate will simply slip into position, do not pin it up tightly at this stage as the strike action will have to be checked.

5. The pin wheel cock provides last minute orientation of the pin wheel and others in the strike train. This part of the work is started by checking once again that the warning wheel pin is directly under the fly when the stop wheel pin is resting on the upper edge of the strike stop arm. If this is not the case, the plates will have to be separated slightly while the adjustments are made. If you pull them only slightly apart, you may well avoid having to reseat all the pivots. It is usually possible to remove the pin wheel completely from the train after removing its cock, making the work considerably easier; the pin wheel is then replaced and positioned when all the others are lined up correctly.

In an idle state, the hammer pin should never rest on any of the pins of the pin wheel. Ensure that this is the case. Replace the back cock.

6. The detent springs may now be replaced. These are adjacent to the arbors of c and e in Fig 8.7. The first

one is for the hammer arbor, which goes under the pin projecting from this arbor. The other is for the strike stop arm; this is fitted in such a way that it pushes the knife edge of the stop piece into the centre of the movement.

7. Finally, the motion work, clicks, and ratchet wheels are replaced, remembering to oil the relevant bearing surfaces.

Making a Final Check

It will now be possible to check that the striking train has been assembled correctly and that it functions in synchronisation with the hands. Give the strike train winding arbor just two winds with a well-fitting key and proceed to make these tests.

1. With the cannon pinion wheel pins clear of the lifting piece, raise the rack gently as far as it will go; it should be held there by the rack hook. If this does not happen, the hook may be mis-aligned on its arbor.

2. Now go back and check that the various wheels in the striking train are in their correct positions: The gathering pallet should face away from the rack at an angle of 180 degrees. The gathering pallet wheel on the other end of this arbor should be locked with its pin resting against the upper edge of the strike stop arm. Normally the pin will rest on the edge that is nearest the centre of the movement.

3. The warning wheel pin should be directly opposite the knife edge that arrests it; this allows the warning wheel one half of a turn when the warning sequence starts. If the warning wheel does not have this run up time, there will be a danger that the striking train will be slow to start, creating an irregular sounding of the bell or gong.

4. Place the minute hand on to the movement and slowly rotate it, making sure that the strike train sounds the hours and half-hour cycles. At the 'five minutes to' position, the warning cycle should start with the warning wheel making a half turn before being caught by the lifting piece knife edge.

5. If you are sure that all is well, replace the dial but push the dial plate pins only a little way in. Move the minute hand round again slowly, making sure that the movement strikes in accordance with the position of the hands.

If any problems are encountered at this stage, refer to the fault finding guide at the end of this chapter. Above all do not be discouraged if the striking is not correct first time; persevere and check all the instructions again.

Reassembling a Locking Plate Movement

Locking plate movements are slightly easier to assemble because there are fewer parts in the striking train and the mechanism is easier to understand.

Replace the components on the front plate using these notes and the illustrations as a guide:

1. Replace the centre wheel after placing a smudge of oil on the front pivot.

2. Now replace the striking parts in the following order (see Fig 8.8): Strike stop wheel *h*; warning wheel *i*; fly *j*; and intermediate wheel *k*. At this stage the pin wheel may be omitted and replaced later when the plates are pinned together.

3. Next replace the locking wheel knife edge *e*, and the hammer arbor below it.

4. On this movement the warning pin will normally need to rest in a different position from that of the rack movement. The pin still needs to be 180 degrees away from the locking plate knife edge, to position it as far away from the fly as possible.

5. The strike stop wheel should have its pin resting on the upper edge of the locking plate knife edge arm.

6. Replace the back plate and the pin wheel, remembering that the cock of the pin wheel may need to be removed first.

7. Now replace the locking plate *f*, taking care to align it correctly with the marks mentioned in the dismantling section.

8. Replace the detent spring for the hammer arbor. This goes on the underside of its associated pin. The locking lever spring is also repositioned with a smudge of oil on the spot where it rubs against the locking lever.

Fig 8.8 Rear view of a locking plate striking movement. The brass back plate has been replaced with one made of Perspex. (e) Knife edge lifting lever, (f) Locking plate, (g) Pin wheel, (h) Pin on strike stop wheel, (i) Warning wheel, (j) Fly, (k) Intermediate wheel.

Making Final Checks

With the plates pinned together it will be possible to check the strike action. Give the strike train spring two winds with a key, and then make the following tests:-

1. Gently lift the locking plate knife once and let it go. The movement will then either start to strike repeatedly or merely once. Now verify the positioning of the pins and wheels. The warning wheel pin should be 180 degrees away from the knife edge of the strike release lever. The hammer pin should be resting between, and not upon, the pins of the pin wheel.

159

2. Replace the dial plate and dial but do not push the dial plate pins right in. Now mount the minute hand and verify the striking against the position of the hands. If the number of hours does not correspond with the position of the hour hand, merely lift the locking plate knife edge and allow the train to run until the correct cam is met on its periphery.

A common problem encountered at this point is that the movement strikes once regardless of the number of hours. This is normally caused by the locking plate being in the wrong position on its arbor, or the lifting pieces and levers of the strike train being out of position. Hopefully, the problem solving guide at the end of this chapter will help overcome any such difficulties.

Setting Up and Testing the Movement

With both rack striking and locking plate movements, I strongly recommend that they be set up dry and tested thoroughly before oiling. There is a good chance that the plates will need to be split to set up the striking and this inevitably leads to the smudging of carefully polished brasswork.

The hours and half-hours should strike when the hands are in their exact positions on the dial. It is for this reason that the dial plate and dial were lightly pinned in position. Replace the hands and give both winding arbors just two turns with a key. Set the movement up on a test stand and proceed to test in the same way as described in the last chapter. If the striking is out of sequence, gently lift the locking plate knife edge and drop it to sound the next hour. Continue to do this until the right number of strikes is obtained. Remember that the longest cam is twelve o'clock, and the shortest two o'clock. One o'clock is a depression.

Oiling

When completely satisfied that the movement is working correctly, complete the oiling. The Swiss company Moebius, mentioned in Chapter Seven, have not published recommendations for oiling these striking movements and commonsense will have to prevail.

Use an oil that matches the recommendation for a similar part in the timepiece oiling chart. For example the fly has a pivot that is almost the same size as the pivots of the pallet arbor, therefore lubricate it with a similar oil. A smudge of oil will be required on all metal to metal surfaces such as the edge of the locking plate knife edge that abuts the locking plate. Be very careful with the strike train. If it is over-oiled, it will repay your generosity by swiftly spraying the rest of the movement. Last but not least, do not forget to place a spot of oil on the middle of the fly spring.

Fault finding

An experienced reader may feel that the obvious has been laboured in this chapter. In my early days, I spent many hours trying to sort out striking problems; the intention is therefore to suggest methods that may avoid much tedious work. I found some books to be helpful, more were confusing, and others assumed far too much prior knowledge. It is virtually impossible to solve striking problems quickly without having a thorough understanding of the movement.

The fault finding table here gives an indication of the likely solutions to some of the more typical problems encountered.

Movement Still in the Case

Occasionally it will be possible to overcome a striking problem while the movement is still mounted in its case.

Clock Stops at Twelve O'clock

The most likely problem with rack movements is that the whole movement stops at or near to twelve o'clock. This is caused by the strike train running down, thereby losing the power to lift the rack tail clear of the hour snail. The going train continues to function until the rack tail comes up·against the high edge of the hour snail. In most cases, winding alone will cure the problem, although a broken or slipped spring can create the same symptoms. Other likely problems are a gummed-up strike train or one that has become damaged.

Clock Does Not Strike

Try not to overlook the obvious. If at striking time you can hear the strike train whirring but nothing is struck,

161

it is possible that the hammer is not striking the bell or gong. The leather tip of the hammer may be missing or the arm may have been bent. Occasionally a hammer head will fall off, only to be found in the recesses of the case.

Locking Plate Striking

If a locking plate movement continues to strike once only, here are some possible reasons:-

1. If the knife edge lifting lever is out of position on its square, it will only allow the movement to strike once before the pin on the stop wheel is caught by the internal arm of the locking plate knife.

2. Make sure that the knife edge lifting lever has not been bent by an inexperienced worker. Only bend it yourself if you are completely sure that this is causing the problem.

3. Check that both pins are mounted on the cannon pinion and they have not been bent at some stage.

4. Similarly, the strike release lever may be bent.

5. The bush that the strike release lever is mounted on may be catching against the front plate of the movement.

Multiple or Non-stop Striking

1. Check that the stop pin on the strike stop wheel is not bent or even missing.

2. The locking plate knife edge may be jammimg in the up position. Check that the knife edge has the detent spring fitted. If so, this may stop it from returning smartly.

3. Ensure that the locking plate is orientated correctly on its arbor. If it is incorrectly positioned on the square, various problems will arise. Try moving it around until the striking is correct.

Clock Does Not Start to Strike

1. Try the obvious: Is the movement wound?

2. Check that the cannon pinion has not moved forward of the line of fall of the lifting piece.

3. Check for misaligned or missing pins on the cannon pinion.

4. The warning wheel pin should always be 180 degrees away from the blade on the lifting piece that stops it during the warning cycle. If the pin has too short a run, the strike sequence may fail to start

Clock Strikes Early

1. If the movement starts striking several minutes before the appointed time, there is a strong possibility that the warning sequence is failing. Warning normally occurs a few minutes before the hours or half-hours.

2. The stop wheel pin may be resting just on the edge of the locking plate knife edge that abuts it. This in turn may cause the striking to start out of sequence intermittently.

Rack striking

1. Verify the basic set-ups. With the movement in the idle state, make sure that the gathering pallet is at an angle of 180 degrees from the rack. The rack should be at the top of its travel with the lowest tooth resting on the rack hook.

2. The pin on the gathering pallet wheel should be resting against the inner knife edge of the strike stop arm. The warning wheel pin needs to be opposite the knife edge of the lifting piece.

Clock Strikes Once Only

1. Read through the notes above for the locking plate movement. Some of them will be relevant here.

2. Some of the worst problems are caused by the rack hook being loose on its square, or out of position.

163

Multiple or Non-stop Striking

1. Check for missing gathering pallet.

2. The rack may have seized in the up position.

A Final Note

Try not to give up if you have a troublesome movement. It is suprising how much may be gained by leaving the work for a day or two and returning to it afresh. Above all, try to resist the temptation to bend parts to make them work because the chances are that one sequence will be forced to work at the expense of another.

IX

Pendulum and Escapement Repairs: Cleaning and Testing the Movement

Fortunately, problems with pendulums are seldom encountered. The biggest headache is probably that of replacing a missing pendulum. If the clock is small, there is a good chance that a simple brass bob would have been employed originally. These are available in a wide variety of shapes and sizes and are surprisingly inexpensive to purchase. Reference to Chapter Five will enable you to select the correct length and, hopefully, a pattern to match the movement.

The pendulums from larger cases could well have been mercury compensated. If the pendulum is missing and it was known to have been of the mercury type, it may be hard to locate a replacement. One supplier in the U.K. is able to produce a unit with twin glass jars for pendulum lengths of 130 mm and 230 mm. If a lathe is available, the construction of a mercury pendulum would make an interesting project, but some research would have to be done to get the sizes correct.

Brocot Pendulum Suspension and Regulator

It is rare to find a French marble clock without either a Brocot or the very similar Vallet suspension, with its totally enclosed brass block.

There are normally only two repairs to be made on the suspension. The first is to replace a damaged spring. The slightest kink in a spring will cause the pendulum to oscillate. Replace it if there is the slightest hint that it may have been damaged. Damage is usually caused by the clock being moved long distances with its pendulum still in position, fortunately this is often all that is wrong with an otherwise good movement. Springs are now available with brass or plastic ends and, although they may look strange, the plastic type are usually better made than the modern brass type.

Both Brocot and Vallet suspension blocks are of simple construction and easy to take apart and re-assemble. Check with the Brocot model that the suspension spring is a good fit between the cheeks of the regulator assembly as excessive play will cause the pendulum to lose impulse, and may even prevent it from working altogether.

The second repair, often required, is to the gear wheel riveted to the end of the regulating arbor or rod. Hold the wheel between your fingers and gently attempt to rotate the road with the other hand. If any slackness is evident, place the rod in a vice that has protected jaws and hammer the rivet until the slackness is taken up. It may be necessary to clean up the rivet with a Swiss file after hammering it.

Silk Suspension

Silk suspensions are a rarity among marble clocks but may be encountered, particularly in other French clocks with similar movements. They were used extensively until 1860, when the Brocot type became universally popular. The theory behind the suspension is simplicity itself, the pendulum length being adjusted by raising or lowering a loop of silk cord. Unfortunately, silk changes its characteristics with both temperature and barometric pressure. After repairing a movement with this suspension do not remove the pendulum again if moving the clock some distance, but simply tie the pendulum to something solid on the movement. Failure to do this will result in the clock needing to be regulated again when it is restarted.

The Crutch and Pallet Arbor

Movements with a Brocot pin-pallet escapement will usually have the Brocot self-setting pendulum crutch. The crutch is attached to the pallet arbor with a friction collar and screw thread. To set the pendulum in beat on an assembled movement, you merely move it gently as far as it will go to one side and then let go. The crutch should automatically take up its optimum position on the arbor and should now be 'in beat'. The only other problem may be looseness of the crutch or the arbor. It is possible to increase the friction by removing the crutch and gently closing up the collet that encloses the thread. In no circumstances should the arbor be soft-soldered to the crutch.

Check to see if there is any undue slackness between the side of the crutch and the pendulum rod. This is best done with the movement fully assembled, but if this was omitted, hold the crutch against the pendulum rod and check for slack. Make sure that the rod is held near the centre part of the crutch and close to the point on the rod where engagement would normally be made.

Fig 9.1 Taking up the wear in a pendulum crutch. The forked edge is held over the squared end of a block of metal or hardwood and closed slightly by tapping with a light hammer.

If there is slackness, place the crutch on a steel stake and carefully spread the brass with a hammer (Fig 9.1). If a stake is not available, try using the head of a large hammer instead. The sides of the crutch must be kept parallel to allow free movement of the pendulum rod after adjustment. When complete, burnish the crutch sides smooth with an oval burnisher or the round part of a small screwdriver blade.

The Brocot Pin Pallet

Brocot pallets are semi-circular and made either of steel or a hard stone such as agate. The jewelled (hardstone) variety is usually encountered on visible escapements. Steel pallets are preferable as they are easy to replace, although they wear very little. Careful inspection of the pallets will reveal whether they are worn by use, or have been stoned by a previous worker. A micrometer is

167

invaluable for checking the dimensions of the pins; the semi-circular width should be exactly half that of the total diameter.

New pallets are available from material dealers in steel or jewel. In order to help select the correct size, some sizes available for Brocot escapement pallets from one particular material dealer are listed here.

Drill Size Number	Diameter
42	2.25 mm
44	2.18 mm
47	1.88 mm
50	1.70 mm
51	1.65 mm

When checking the size of the pallets, or making them, it is preferable to use a slightly smaller diameter rather than a larger one. The escapement will function with undersized pins, but may not with oversize.

The pallets are usually held in position with shellac, and may be removed by soaking in methylated spirits. Alternatively, gentle heat will soften the shellac. In visible escapements the pallet stones are nearly always a sloppy fit in their holes, making it very difficult to align them correctly. My method is to bush the holes until the stones just fit; the escapement may then be adjusted. When correctly positioned, the stones may be shellaced in place.

The Anchor Escapement

The recoil or anchor escapement first appeared during the latter half of the seventeenth century. The most common form found in these movements is the English anchor. The only problem commonly encountered is worn pallet faces. They may be re-aligned to provide an unworn face by moving the pallet piece sideways along the arbor. Many restorers find it acceptable to soft solder a piece of mainspring on the pallet faces as it is almost impossible to close the pallets without splitting.

Platform Escapements

The platform escapement (Fig 4.15) is occasionally encountered. It is easy to ruin a platform escapement if you do not know what you are doing. Experienced readers will know how to clean them; others should leave the task to an expert. It is possible to buy new platforms and their fitting is really only a matter of commonsense. Before ordering a new one, it will be necessary to calculate the number of vibrations, as detailed in Chapter Five.

Adjusting Pallet Arbor Drop

On the front plate of each movement is located a pivot for the pallet arbor, mounted in a turntable. Movement of the turntable will correct the drop, but it is often used to assist in depthing the escapement. Drop is the distance the pallet moves when freed, before being arrested by a tooth of the escape wheel.

Repairing Screws and Threads

A damaged or missing screw may pose considerable problems during restoration. The usual type to work loose are the shouldered screws that hold the winding clicks in position, and these are used here as an example when overcoming most screw problems. This screw is subjected to a constant sideways force from the click wheel, which is occasionally interrupted by the mainspring being wound.

Click screws will often slowly work loose, and, once loose, start to open up the threaded hole in the front plate. The simplest solution is to bush the hole with a clock bush or bouchon and then re-tap it. Unfortunately, the threads of the movement will not always match their modern equivalent because modern metric threads are usually metric coarse. A degree of trial and error will be required here.

Thread Locking Fluids

Modern thread locking fluids sold under the brand name of Loctite or Threadlock have proved invaluable for securing some slightly damaged screws such as the click screws. Ensure that the fluid is of the releasable variety. All threads must be spotlessly clean and use only a tiny drop of the fluid.

Rusted-in screws can be eased by heating them with the pencil flame of a gas torch or the tip of a large soldering iron. Alternatively a drop of paraffin (kerosene) will often act as a penetrating oil.

169

To polish and blue a screw head, hold the threaded part lightly in a pin vice and spin this with one hand whilst rubbing the head with an emery stick. Damaged screw slots are tidied up with either a Swiss or a slotting file. The head may now be blued either in blueing salts or held in a low flame until a good shade of blue is achieved. Plunge it immediately into some oil to quench the metal and give it a protective coating.

Cleaning the Movement

Cleaning should not be started until all repairs have been completed and the movement stripped down.

Of the many ways of cleaning I have used, the following method has been successful for several years. The main benefit is that neither solvent nor alcohol are used, an advantage when considering both safety and cost. A litre of the diluted cleaning solution described earlier in Chapter Seven will be needed, with two plastic containers having a capacity of approximately two litres each. Plastic ice cream tubs are excellent. If undertaking repairs on a regular basis, it may be worth investing in an ultrasonic bath; although fairly expensive, it will enable some of the more inaccessible parts of the movement to be cleaned quickly.

First put on some rubber gloves and then pour a litre of cleaning solution into one of the containers. When fresh, it will give off acrid ammonia fumes; these are not particularly dangerous but are unpleasant. For this reason, try to do the cleaning near an open window.

Cleaning the Plates

Start by cleaning the two clock plates. Immerse them in the solution for about ten minutes, giving an occasional brush with a natural bristle toothbrush. Ensure that all parts are covered all of the time, and try to keep items in the solution for the shortest time possible. If the plates are partly immersed a water line mark will develop. The plates clean up fairly well, but will usually have some staining on them that can be removed later with polish.

When satisfied with the results, put them into a second container which should be filled with one and half litres of hot water with just one drop of washing-up liquid added. This helps clean off the ammonia solution,

170

but more importantly acts as a wetting agent that inhibits the formation of water droplets during the drying stage. This in turn is helped by the residual heat in the metal components. Using more than one drop of detergent in the water will leave a film on the brass work that could later play havoc with lubricants.

Cleaning Component Parts

While the plates are in the hot water, continue to clean the other brass parts of the movement. Remove the mainspring from its barrel because this should not be cleaned in the solution. Great care will be needed with wheels and pinions, ensuring that each individual tooth is cleaned, by brushing across the wheel with a clean but wet brush. The crossed out sections (spokes) of wheels are cleaned with a small piece of pegwood at a later stage.

When all parts are clean and soaking in the water bath, dip a gloved hand into the cleaner and check that some small part has not been missed. Next, wash the rubber gloves under running water, and proceed to remove parts from the water bath. They may be laid out on a towel on top of a boiler. The object is to dry them quickly before the evaporating water can leave any stains on the bright work. Take each piece in turn, drying it first with soft tissue, and then buff it with the soft clock brush. Occasionally wipe the brush across a piece of clean chalk to help keep it fresh and free of dirt. After chalk brushing the pinions, clean them with a wedge-shaped piece of pegwood.

If the arbors have become rusted, they can be cleaned with a rag and metal polish. Severe rusting should be rubbed gently with fine emery paper glued to a flat stick; spinning the arbors in a pin vice will help. In order to preserve the bright steel work from rust, rub all surfaces with a lint-free cloth that has been moistened - not saturated - with a drop of clock oil.

Take a sharpened piece of pegwood and use it to clean out all the pivot holes in both the back and front plates. Keep scraping the wood clean with the edge of a sharp knife and twirl the wood in the holes until it comes out completely.

Clean the oil sinks with pegwood dipped in metal polish, or with a fibreglass pencil. If the sinks are left dirty, there is a danger that they will not retain the oil when it is applied.

By now it will probably be appreciated why it costs so much to have French movements cleaned professionally.

When completely satisfied that everything is bright and clean, polish the front and back plates with a long term type of brass polish, making sure to buff up and down only, from the top of the plate to the bottom. A circular cleaning motion will leave a strange finish on the plate. Remember also to clean all the brass parts that attach to the plates, such as the cocks and Brocot suspension blocks.

From this stage, try not to touch exposed metal surfaces with your fingers. Hands exude moisture that is slightly acid, and in time this will rust steel and mark brass. There are several ways to overcome the problem. The simplest is to dust your hands occasionally with talcum powder or to use the small rubber finger covers that are used by watch and clock factories to overcome the marking problem. Holding wheels and plates by their edges will also help.

Re-Assembling the Movement

If you are not used to working on clocks, take reassembly at a leisurely pace. Do not hurry, because it is easy to make a mistake when tired or rushing to get everything back together and working.

The mainspring barrel may be polished after the spring is re-loaded. Before hooking the spring on to the winder, check once again that the barrel hook and spring are in good order. Hold the mainspring arbor up to the barrel hole and make a note of the position of the arbor hook, which will enable you to position the spring the correct way round. The arbor hook will pull towards the small end of the spring. This may sound obvious but is easy to do wrongly.

It is possible to re-load without a winder, although it is also possible to distort a perfectly good spring by hand loading. With the spring in the barrel, replace the arbor in the barrel. A few turns with a key will verify that it has hooked. Apply two, and only two, drops of oil to the spring. An excess of oil applied here will only creep out on to the plates and ruin all the careful cleaning work that was undertaken earlier. Alternatively, the barrel and spring may be lubricated with the special liquid mentioned at the end of Chapter Seven.

Take the barrel cover and offer it up to the barrel, making sure that the slot is lined up with the mark on the barrel tube. The cover will go in halfway, and must then be tapped in with a block of wood, or the handle of a brush (Fig 9.2).

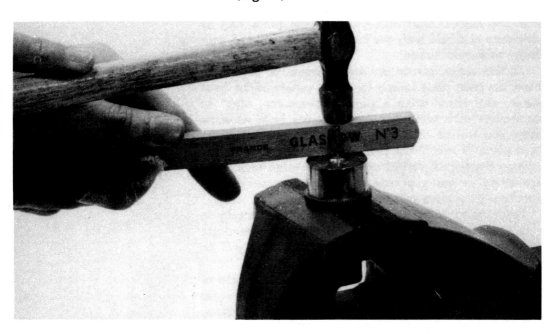

Fig 9.2 Replacing a barrel cover. One or more taps with a light hammer on a piece of wood held across it, will cause it to snap into place.

The Going Train

Take the small cock that fits on to the back plate and position it with its relevant screw. If the screws are mixed up, look for a punch mark on one that corresponds with a similar mark on the cock.

Assemble the going train by starting with the centre wheel. This is the wheel with the long arbor on which the minute hand is mounted. Put a small drop of oil on the front pivot, and then insert it into the plate. The plate is best placed downwards on an ashtray or small box that is covered with material to protect its brass surface.

Next take the barrel, put a small amount of oil on the arbor pivots and place it in position, followed by the rest of the wheels. If all goes well, the escape wheel should have been last. Do not oil the sinks of these wheels as this will be done later, when the movement is completely assembled.

173

Placing the Back Plate

The next job is the challenging part of the assembly work, replacing the back plate. The safest way to do this is to leave the front plate resting on the box or ashtray and then move your head down until your eyes are level with the plates. Alternatively, the front plate may be picked up and held with one hand, suitably covered with finger rubbers or tissue.

At this stage, pivots can easily break, so take care. Place the back plate loosely on to the pillars of the front plate, and then, with a pair of tweezers, start to manoeuvre each pivot into its particular hole. At some stage you will be able to apply gentle pressure on the back plate. It is difficult to describe this process. If you become frustrated, reflect on the knowledge that there are twice as many pivots on a striking movement! Eventually the plate will snap into position.

Before continuing, check that each pivot is actually in a hole, then very gently rock one of the intermediate wheels. If there are any meshing problems it will be jammed solid, indicating a fault in the train.

Still holding the movement, take a steel taper pin of the correct size and gently push it into a pillar hole. Do this with the other pillars and then go back round them, removing and trimming any excessively long pins to length; finally, round any sharp ends with a file. There may be a temptation to use brass pins as they are more attractive, but brass pins are brittle and liable to snap, leaving the tricky job of boring out the broken bits. Be warned, therefore, and always use new steel taper pins throughout. A clean movement held together with mangled pins is unsightly.

When satisfied that everything is in order, gently tap each of the pins once with a light hammer to secure them.

Motion Work, Click Work, and Escapement

Place a smudge of oil on the back of the ratchet wheel, then mount it over the barrel arbor and secure it with its brass cock. Another small amount of oil goes on to the shoulder of the click screw spring, placing a little oil where it rubs against the click. Similarly, oil the tooth of the click itself.

The motion work that drives the hands is mounted next, starting with the cannon pinion and pushing it firmly on to the centre wheel arbor. Follow this with the minute wheel and its cock, then the biggest wheel, the hour wheel.

Take the pallet arbor and place a tiny amount of light oil on the rear pivot, then offer the pivot block up to the arbor. Take extra care because the pivots on the pallet arbor are the finest in the movement. When you are satisfied that they are correctly placed, insert the screw that holds the whole block assembly to the back plate.

The next step is to test the escapement and adjust it if necessary.

Fig 9.3 A movement test stand. Supplied by Southern Watch and Clock Supplies, Orpington, UK.

Setting up and Testing the Movement

The barrel shape of a French movement can be a nuisance when trying to find a suitable place to rest it during testing. With the dial and front bezel in position, the whole movement becomes unstable and is sometimes in danger of toppling over on a poorly-designed stand.

It is possible to buy specially made stands for French movements (Fig 9.3) although a vice will usually serve the purpose well.

To test, run the movement dry for a while before oiling the sinks that were omitted during assembly. This is particularly useful with striking clocks where it may sometimes be necessary to split the plates several times in order to set up the striking train. Non-oiling at this stage will prevent the plates from becoming smudged with oil. It has the added advantage of demonstrating that a movement will run without help from oil.

Still taking care not to get fingermarks on the plates, put the movement on the test stand and hook the pendulum on the suspension. Adjust the movement so that the pendulum is hanging down in line with the central wheel pivot hole in the middle of the back plate. This will ensure that the movement is aligned vertically. Give the mainspring two turns with a good fitting key, and then give the pendulum a push to start it going.

Putting the Movement in Beat

On non self-starting movements, rotate the movement slightly in either direction until a steady, even, tick is heard. The pendulum will need to be swung to re-start the movement if it stops during the 'setting-up' process.

If the pallet arbor is threaded, the clock has the self-setting Brocot escapement and all that should be required is to move the pendulum one way until a slight resistance is felt, then release it. The crutch should automatically set itself in beat. With luck the movement will run for several days with one small winding. If problems are encountered refer to the following list of possible causes. It does not attempt to be exhaustive, but should provide some clues.

Fault finding:

Cures for Stopping and Poor Timekeeping

Always attempt to analyse a fault before dismantling the movement.

Dial

1. Hands locking together.
2. Hand touching dial or bezel glass.

Motion Work

1. Minute hand collet binding against that of the hour hand.
2. Cannon pinion loose on centre wheel arbor.
3. Hour wheel catching against the minute wheel cock.

Pendulum and Suspension

Most problems here are associated with the pendulum wobbling or lacking impulse.

1. Movement not aligned correctly, causing the pendulum to go out of beat. Rotate movement slightly in either direction until a steady even tick is obtained.

2. Similar to 1, but if out of horizontal truth, the pendulum will rub against the back of the crutch, or cause the suspension spring to be out of alignment with the suspension block.

3. Slackness in cheeks of crutch.

4. Crutch cheeks too tight.

5. Pendulum bob loose on rod.

6. Pendulum hook not sitting snug on suspension spring pin.

7. Wrong pendulum fitted:- Too short will make clock excessively fast; too long will make it slow.

8. Movement loose in case, causing it to move out of truth as it is wound, or in bad cases, to rock with the movement of the pendulum.

9. With Brocot suspension blocks, the spring may be loose in Brocot cheeks, causing the pendulum to oscillate.

10. On Brocot self-adjusting crutches, the crutch may be too slack or too tight on the pallet arbor.

11. Suspension spring too stiff causing lack of impulse to pendulum.

12. Suspension spring too wide for Brocot block.

13. Suspension spring twisted or broken.

14. Suspension spring leaves are of different lengths.

Escapement

1. Depthing of escapement pallets incorrect.
2. Brocot pallets too large.
3. Geometry of pallets incorrect.
4. Escape wheel teeth damaged.

5. Excessive wear in pallet arbor pivot holes, usually the rear one.

6. Re-bushed pallet arbor holes out of truth causing the arbor pivot to bind.

7. Worn pallets on an anchor escapement.

8. On a clock fitted with a platform escapement, check depthing between the contrate wheel and the escape wheel.

Going Train

1. Dirt between pinion leaves.

2. Excessive wear in pivot holes upsetting he depthing.

Mainspring and Barrel

1. Barrel teeth bent or broken, usually due to overwinding.

2. Lack of shake (side to side movement) between barrel pivots and movement plates.

3. If during winding the key rotates loosely, either the mainspring has broken or it is not hooking with the barrel hook or winding arbor.

4. If arbor is not clicking during winding, check for missing/loose clickwork.

5. Barrel cover hole worn.

6. Spring oversize for barrel.

General

1. Using too heavy an oil on the finer pivots may cause them to freeze intermittently, especially during cold weather. Try keeping the clock in a warmer place.

2. A useful way of finding a fault is to time how often the movement stops. If it always stops at the same time or time interval, this may indicate a damaged tooth on a wheel, or a leaf on a pivot.

Do not forget that some imitation French marble clocks have 30-hour movements.

Finally, do not despair if you are unable to locate the problem. Leave the clock alone for a while and take a fresh approach at a later date. You gain experience from the movements that do not work, not the ones that are merely cleaned and worked first time.

Oiling

Lubricants were described in the materials section of Chapter Seven. Now that the movement is working in the dry condition, restoration work is completed by oiling the movement. This is a subject often misunderstood by the layman who does not usually realise that large amounts of oil are more likely to make a clock stop than to work.

Remember that a light oil is used for the fine pivots and a heavier grade for the larger pivots of the centre wheel and barrel.

Oil sinks were devised towards the end of the eighteenth century to prevent oil creeping away from the pivot on to the plate. The oil is applied to the sink with an oiler or with a special tool that looks rather like a fountain pen. Just enough oil should be applied to form a meniscus of oil at the bottom of the sink. The sink should NOT be filled to the brim with oil.

Oil poured into an oil cup before oiling, should always be covered when not in use. Each time the clock oiler has been used, it should be either dug into pithwood or cleaned on a lint-free cloth.

X

Styles of Clock Case: Marble or Slate?

The design of a French marble clock case varies from a strong ornamental appearance through to a clean-cut almost austere simplicity. There is not one single design influence that may be traced, making any form of analysis difficult. A glance at pages 223 to 243 will show the wide variations in shapes and styles encountered.

The logical way to approach the subject would be to trace the origins of the designs back to the beginning of the nineteenth century when they first appeared. I once attempted to do this and despite much research was unable to find a particular period from which the designs evolved. I have therefore taken five typical clocks and looked at the influences that may have taken part in their creation. It is stressed that this is my personal interpretation and others may have different ideas.

There are five main designs of case: Plain, Break-Arch, Drum, Greek Temple, and Egyptian.

Fig 10.1 A plain case that owes its simplicity partly to economies of production.

The Plain Case

The simple case illustrated in Fig 10.1 is particularly common and owes its shape more to the economies of its manufacture than to any aesthetic ideal. The construction is mainly of slabs of marble that have been bonded on to a plaster carcase. Only one curved moulding is in evidence, above the plinth. Marble was an expensive material to both quarry and shape and a simple moulding such as we see on this clock would either have been

Fig 10.2 A plain case with an elegant design.

Fig 10.3 Another plain case that is typical of many French marble clock cases.

carved by hand in the early days, or produced on an expensive machine towards the end of the nineteenth century. The simplicity of this design is therefore a function of the cost of manufacture as opposed to any particular style.

Bearing in mind that the simple marble case was cheaper to produce, we can see why many marble clocks conform to the shape seen in Fig 10.2 and Fig 10.3. With all these clocks, ease of manufacture was a predominant factor rather than any artistic or aesthetic influence.

The Break-arch

The fine clock in Fig 10.4 has a fairly plain shape but has the added feature of an arched top. This appears to be a type commonly known as a 'break-arch', often encountered on bracket and longcase clocks, in which the arch top is slightly narrower than the case to which it is joined by short shoulders. In fact the arch only appears to be 'broken' by the ledge on the outside of the case.

The case shown in Fig 10.4 would have been fairly expensive to produce because of the rounded top and mouldings that surround the bevel-edged glass front panel. The fine quality is enhanced by the use of a Brocot visible escapement and a mercurial pendulum.

Fig 10.4 Marble case with an arched top which looks like a break arch, a relatively simple design, but with the arch adding to the cost of manufacture.

The Drum

The drum case design is one of the most attractive and appears in all shapes and sizes. The model shown in Fig 10.5 is of a medium size that would look attractive in many modern homes. The three main parts of this case – the drum shaped top, the middle section and the plinth – are worthy of examination.

The drum shape of the marble surrounding the movement, which gives the case its name, is a fine example of a marble mason's craft and would have required various stages of machining and finishing in its production. The drum would have been turned on a lathe, and then incised to form the grooves or 'volutes' on its outer surface. These no doubt were inspired by classical Greek columns; another example of the mixture of design influences that were involved.

Fig 10.5 The drum style of case, named after the large drum encircling the movement. Note the volutes (incised grooves) .

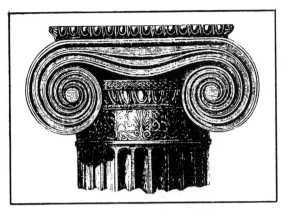

Fig. 10.6 The capital of an Ionic column. Note the similarity between this, if turned upside down, and the support of the drum in fig 10.7.

The centre section of the case which supports the drum is often made of an ornately carved piece of marble that would almost certainly have been fashioned by hand. Marble carving machines were available in the nineteenth century but it is almost certain that they would have been unable to perform this sort of delicate work. The shape of this support is reminiscent of the spirals found in the Greek Ionic period; compare it with the capital or headpiece of a Greek Ionic column in Fig 10.6. The volutes on the upper surface help to break up the solidity of the carving.

The plinth is made up from laminated pieces of marble that have been grooved to reduce their solidity and a coloured stone 'sandwich' serves to break up the uniform blackness of the case.

A simpler version of the common drum case is shown in Fig 10.7. Here the marble carving has been kept to a minimum and the absence of volutes tends to make the drum look rather solid and clumsy. The marble in this clock is of a poorer quality than that in Fig 10.5 and tends to suggest that simplicity of design often went hand in hand with an economically produced case.

The fourth type of case is categorised as those more influenced by classical Greek art and architecture.

Greek Temples

The ancient Greek empire had an immense effect on the art and design of the nineteenth century in both France and England. Traditional sculptors followed the classical model during the eighteenth century and after the French Revolution many others in the fields of art and architecture turned to ancient Greece for inspiration.

The marble case often associated with the Greek influence is similar to the example in Fig 10.8. The long columns are modelled on the temples of Ancient Greece and form a style that, particularly with a roof of triangular profile, was particularly popular at the turn of the nineteenth century. The Manchester catalogue of Hirst Bros, circa 1910, shows twenty-six marble, or imitation marble, clocks with a strong Greek temple influence.

185

Fig 10.7 *The simplest of drum cases, and consequently least expensive, to make.*

Characters from Greek mythology have also provided artists and architects with inspiration for many centuries. Although we do not often see human beings incorporated in French marble clocks, there are occasions when a Greek or possibly Roman figure is employed. An example of such a case may be found in Collectors' Clocks, Fig 29, at the end of this book.

Egyptian Cases

An Egyptian style case is illustrated in Fig 2 of Collectors' Clocks. Egypt exerted a powerful influence on both nineteenth century England and France. Napoleon's expedition to Egypt in 1798, the Battle of the Pyramids, and then Nelson's defeat of the French fleet in the Battle of the Nile, fired the imagination of the British public.

186

Fig 10.8 This clock has been dated circa 1885 and marks the beginning of the Greek column period. After this date, more and more cases were produced using these columns. By the 1900s most marble clocks were of the Greek temple variety.

Napoleon's expedition was accompanied by 167 members of a Commission of Arts and Sciences consisting of engineers, cartographers, and specialists in many other branches of the natural and physical sciences.

Public interest in ancient Egypt reached a peak in 1820 when the Pasha of Egypt welcomed trade with

187

Europe, and particularly Britain, where painters and poets, such as Turner, Keats and Byron, began to exhibit an Egyptian influence in their work. The Great Exhibition of 1862 in the Crystal Palace, enraptured the public with an exhibit of the Egyptian Court and a display of Egyptian antiquities from the Cairo Museum.

Fig 2 (Collectors' Clocks) shows a fine example of a French marble clock that is unmistakably Egyptian in design. The clock was made around 1880 and the dial is flanked by pillars, or 'pilasters' as they should be called, with Egyptian masks on the top of them. The panels on the case are inscribed with hieroglyphics and charioteers, while a statuette of Cleopatra seated on a Sphinx rests regally on the top of the case. On each side of the case stands a similarly decorated bronze and marble urn.

Many French marble clocks were made in this style, often presenting a strong and bold design that cannot fail to capture the eye. When properly restored, they represent some of the finest examples of French marble clock design.

A Pastiche of Styles

There are French marble clock cases where, at best, a mixture of design influences prevails. This is not unusual. Today motifs are still borrowed from a broad range of historical influences. A study of these helps with the appreciation of the French marble clock, and has the added benefit of providing interest when walking around cities such as London, when one can make mental comparisons of the similarities between some of the beautiful old buildings and the humble French marble clock.

The Nature of Marble

In everyday conversation, the word 'marble' is used by laymen, dealers, and quarrymen, to describe several types of stone that have the same general appearance. The term was put in its place by W. G. Renwick, in his book 'Marble and Marble Working' published in London in 1909:

"The term MARBLE is one that is variously understood. From the scientist's standpoint only metamorphosed limestone should be included in its definition.

Commercial men however are seldom scientists during business hours especially if science interferes with their turnover, although a strict interpretation may limit the term to crystalline calcareous and magnesian formations. From a commercial standpoint MARBLE is recognised as including any natural stone that is of less hardness than granite, having a sufficently close texture to take and retain a polished face and being produced in such quantity as to be available for use for decorative purposes."

It is common knowledge that much of the stone used in the construction of French marble clock cases came from Belgium, although there has always been some doubt as to whether this stone was a true marble. At various times the clocks have been called 'marble clocks', 'Belgian slate clocks', 'Belgian limestone clocks', or merely 'slate clocks'. During my research Gerald Culliford, a marble importer based in Kingston-upon-Thames, England, immediately identified the material as Belgian black marble; he stocked it and was familiar with its use in clock case making, being the owner of a French marble clock made of Belgian black.

A second opinion was given by Allan Jobbins, a specialist in minerals and gemstones who was Curator of Minerals and Gems at the Geological Museum, London, from 1950 to 1984. He too identified the stone as Belgian Black, but added the following,

"Belgian Black is more correctly described as limestone in the true petrological sense of the word". He also agreed with my suggestion that it would be rather pedantic to refer to the material as 'Belgian limestone.'

In 1882, a Belgian specialist called G. Dewalque* created the terms 'limestone marble or calcareous marble', depending on the translation, to describe all limestones, metamorphosed or otherwise, that could be worked with the same industrial processes that were used with the true marbles.

There is no doubt that most cases were made out of Belgian Black marble although they are commonly referred to as 'French marble clocks'. If asked whether the stone is a true marble or a limestone I say, "They are made of a calcareous marble", a term that I believe should be applied to most of the Belgian black marbles. After all, the description was evolved by Belgians who have had considerable expertise in this area for a long time.

Footnote:

*".... G Dewalque.... 1882.... avait crée le terme 'Calcaire-marbre' pour designer tout calcaire recristallise ou non, a grain discernable ou a texture aphantique, qui peut etre industriellement traite oomme marbre." -- L'industrie du Marbre en Belgique (E. Groessens) 1981, Belgian Geological Department.

The term 'calcareous marble' can be applied to non-Belgian marbles if their geological formation is of a similar nature. The subject of specific Belgian marbles is discussed later in this chapter.

Fig. 10.9 Testing for a true marble with acid by pouring a little vinegar on a part of the case that is normally hidden.

The Acid Test for Marble

It can be difficult for the layman to determine by sight whether a case is made of slate or marble; I have devised a simple test that will determine the material used. Marble has one property that differentiates it from slate; it effervesces in contact with acid. Slate will not react in such an obvious way. Remember that British slate roofs have survived many years of exposure to the acid atmospheres of coal fired Victorian cities.

The main materials required for testing are a small quantity of ordinary vinegar (acetic acid), some glasspaper or emery cloth, and a watchmaker's eyeglass. Take the case that is to be tested and ensure that the pendulum has been removed. With the abrasive paper, rub a section of stone on the underside of the case; a small strip about two centimetres long will do.

190

Make sure that the area is cleaned back to virgin stone and that it is not contaminated with old wax or dirt. Do not blow the dust away but leave it to settle on the bare strip. Place a drop of vinegar on the clean patch (Fig 10.9), then watch it carefully with the eyeglass. If the material is marble, it will fizz slightly. The bubbles will be quite tiny and are not always visible to the naked eye. When the test is complete, wipe the bare stone with a damp rag to kill the effect of the vinegar. Other acids may work just as well, but vinegar is the safest and easiest to obtain and use.

While you have the eyeglass to hand take a close look at an area of the case that has become rubbed bare of wax or enamel. Sometimes you will see small gold flecks on the surface. These are particles of pyrites, popularly called 'fool's gold', a sulphide mineral that is occasionally found in slate or limestone.

The acid test is simple to perform and does not damage the case in any way. You can even use it on a case that has been heavily lacquered, provided that the sample area on the base is clean. It is possible that a clock dealer would allow you to undertake this test, especially if you are parting with a considerable sum of money for the clock.

The Nature of Marble

The chemical content of a true marble changes significantly from one sample to another. With the marbles known as 'geological', the basic constituent is nearly always calcium, which combines with carbon dioxide and oxygen to form the mineral calcite ($CaCO_3$). This is made up of 44% carbon dioxide and 56% lime, or the rock limestone.

Subjected during great geological changes to extreme heat and pressure, the limestone slowly altered physically while retaining its chemical properties. The final material is a metamorphosed rock that is very crystalline in structure and of greater density than the original. Belgian Black is not a true marble because it has not been subjected to this process of heat and pressure.

Marbles (true and otherwise) are classified under eight separate headings. The names used give an indication of their characters.

191

1. Saccharoidal. These have even grain and look sugary when fractured. Some of the finest statuary marbles come under this heading. (Sacchar is Greek for sugar).

2. Unicoloured. An evenly coloured marble without veins or obvious markings. The unicoloured marbles are usually sound, obtainable in large pieces, and weather well. For architectural work they were frequently used with Sicillian marbles to produce marble floors. Included in this category are the blacks of which Belgium black is probably the best known.

3. Variegated. These have patches and veins of different colours distributed randomly throughout them. Their lack of strength led to them being used mainly for lining walls.

4. Fossiliferous. Marbles that contain fossil remains in one shape or form and are among the oldest of the rock formations. Fossiliferous marbles are usually found in Great Britain, Belgium, Brittany and the north of France.

5. Crinoidal. The crinoidal marbles are composed of fossilised shell fragments. They are fairly sound but few were used for external work.

6. Brecciated. These are composed of angular particles of limestone with various colours and formed from pieces of older rock that were crushed by other physical forces.

7. Laminated. Regular veining and bands of different colour will indicate a laminated marble, caused by the original formation of the marble.

8. Serpentinous. Those marbles that are wholly or partly composed of the mineral serpentine. Most of the dark green marbles come into this category.

Black Marble

Chemically pure marbles are not found in Nature; all have varying quantities of substances in them that cause the familiar discolourations. Some of the more common colourants are carbonate of magnesia, carbonaceous matter, mica, talc, and the carbonates and oxides of iron, silica and clay. Note the term 'carbonaceous', which implies that the material contains carbon that gives black marble its characteristic hue. Black marbles such as Belgian black are often quarried near coal mines. The black marbles come from the unicoloured group; they were usually sound when mined, available in large sections, and weathered well.

Onyx Marble and Onyx

True onyx is one of the chalcedonies, a group of hard ornamental stones which includes agate, jasper, and sardonyx. Onyx marble is not so hard, and neither a true onyx nor a true marble, although it is commercially considered to be marble. Onyx marbles are banded calcite or travertine (hard and semi-crystalline stones deposited from water holding lime in solution). They are of two kinds both composed mainly of carbonate of lime.

One kind will have been produced by the action of cold water on the insides of limestone caves, forming pockets of marble, while the other is precipitated from the lime found in the basins associated with hot water springs. Onyx is coloured by metallic oxides.

Alabaster and the Decorative Marbles

Alabaster is a light cream and occasionally white type of gypsum. Gypsum is calcium sulphate ($CaSo_4 + 2H_2O$), and is composed of 46.5% sulphur tri-oxide, 32.6% lime, and 20.9% water. It is really too soft for cases although very occasionally used.

A number of coloured marbles were used to decorate the clock cases. Some were quite expensive and led in some instances to the use of substitutes.

In 1889 an English writer, Julien Tripplin[81], noted the various marbles employed at the time: "Marble clocks require more than one colour to make them attractive; white and grisotte marbles are drawn from the Pyrenees, red from Greece, malachite and lapis lazuli from Russia, onyx from Mexico".

81 Tripplin.
Watch and Clockmaking at the Paris Exhibition of 1889: Being an Account and Comparison of the Exhibits in the Horological section of the French International Exhibition. London 1889.

XI

Extraction and
Types of Marble

The end of the nineteenth century saw tremendous progress in the technology of quarrying. The Americans were probably the most advanced in terms of efficency followed closely by the Belgians and the French. The English lagged behind and never really turned marble quarrying into a particularly profitable exercise. Electricity, steam, and compressed air played an important role in the mechanisation and automation of the quarries and mines.

There were five main methods of extraction: Mining, blasting, wedging, sawing, and channelling.

Mining was used both in England and Belgium as a means of extracting some of the better quality black marbles. A mine in England was opened in 1832 at the Rookery Plantation near Ashford-in-the-Water. The first Belgian mine opened slightly later in 1859 at Mazy. The nature of the rock formation meant that the strata were inclined at an angle, making open quarrying inefficent.

Blasting, the crudest method of extraction, was the principle method employed in England. One limestone quarry in Devonshire produced both road stone and block marble. If the blocks produced by blasting were big enough, they were sold for decorative purposes; otherwise the stone was used for roadmaking.

Using explosives is wasteful because large quantities of loose material is produced with each blast. The explosion also unsettles the structure of the stone, sometimes causing it to split or disintegrate when worked at a later date. Black marble is an exception as it can usually withstand the shock of a blast, no doubt owing to its hard and close-grained structure.

Wedging is the oldest and simplest method of extraction. Steel wedges are driven between the veins and rifts that naturally occur in a marble formation. It is an extremely wasteful method because marble rarely occurs in regular veins and the blocks produced by this method are of many shapes and sizes. However, the process is successful with slate, which has a regular formation.

Occasionally drills and wedges were used together. Two men formed a team with one holding the drill while the other struck it with a heavy hammer. A long wedge was inserted after the hole had been drilled between about two-thirds through a block. Naturally the number of holes drilled varied according to the size of the block.

The hand drills were superseded at the turn of the century by machine drills powered by steam, air, and sometimes electricity, Fig 11.1. Many machine drills were of American manufacture although British machines were available that matched the quality of the American tools. Machine drills were used in France and Belgium although their use was by no means as widespread as the traditional hand drilling technique.

The Wire Saw

The wire saw, Fig 11.2, was widely employed in both France and Belgium from around 1880 onwards. It is an endless wire rope with three strands, the length of the saw depending upon the size of the work in hand. The wire was passed round a central driving wheel of about 1.2 metres diameter from which it was carried on pulleys around the quarry at a speed of between 12 and 14 miles per hour. The pulleys varied between 150 mm and 600 mm in diameter depending on whether the pulley

THE SULLIVAN ROCK DRILL.

Fig. 11.1 The Sullivan pneumatic rock drill.

Fig. 11.2 The wire saw. A continuous loop of wire cuts through large sections of stone.

196

was acting on the wire in a straight line, or if a sharp bend was being negotiated.

The wire saw was invented by Eugene Chevallier and patented on April 8th, 1854. The system remained unused for some twenty-five years until it was revived in 1880 by Paul Gay. Later developments were introduced in 1884 by the Belgian, Michel Thonar. His system used universal joints which with several other improvements made the saw into a flexible and versatile tool.

When a vertical cut had to made into a deposit, large vertical shafts for the wire saw to run in, were initially cut with a perforating drill. Fig 11.3 shows a complete marble core being lifted out of the ground by one of these machines.

Fig 11.3 The perforating drill. This massive cutting tool was used for making the initial hole for a wire saw. A complete column of marble could be cut and withdrawn in one operation.

The Penetrating Pulley

In 1898 the wire saw was considerably improved by the introduction of the penetrating pulley. Fig 11.4 shows a

197

pulley mounted on the bottom end of an upright shaft. The pulley wheel is slightly thinner than the sawing wire, allowing the pulley to enter the slot after the wire saw. Sand and water were fed to the wire to provide the cutting action.

Channelling was used extensively in the USA where the larger market warranted its use. The machines employed were highly advanced for their time, often being powered by electricity and producing large quantities of marble involving little manpower with very little wastage of material.

Marble Production in France

The production of marble from Gaul predates the French nation. The Romans quarried and worked marble there and at least four Gallic marbles are known to have been used for the decoration of Imperial Rome. In later times the Kings of France were to encourage the industry, especially during the Renaissance. From Louis XII onwards, systematic stockpiling of marble was undertaken, culminating in the construction of the Palace at Versailles under the influence of Louis XIV.

The industry almost collapsed during the unrest created by the First Republic and the Napoleonic regime and it was not until 1835 that any visible improvement in its fortunes became apparent. Progress was slow after this date, but led to the majority of the older marbles being quarried again. In 1887, Arthur Lee, in his book 'Marble and Marble Workers', recorded one hundred types of marble that were mainly being used commercially. Of these, at least thirty-two were of French origin.

In the same year of 1887, the French imposed heavy import duties on Belgian marble. By hitting the Belgians hard, this must have boosted the production of marble quarried on French soil as intended. This implies that much of the marble used by Japy Frères (see Chapter Two) came from France. Japy had a reputation for efficient and economical working and would not have imported from Belgium if the import duties were high.

Fig 11.4 The penetrating pulley, an improvement on the wire saw. The pulley was narrower than the wire, allowing it to penetrate a surface as shown here.

At the turn of this century the French were using a combination of the wire saw coupled with drilling and wedging. Blasting was only performed in exceptional conditions. An experiment with channelling equipment was tried in a few places but the results did not justify its long term use. Probably mining was used.

Fig 11.5 Machine tool for facing blocks of marble.

Marble Working Companies

There is little doubt that the bulk of all marble clock cases came from Belgium. Although the French industry was by no means as large as that in neighbouring Belgium, the work was still considered by some authorities to be of high quality. The only criticism made by Renwick[72] in 1909 was that the French were "better at promising than performing and delays are often encountered which might with better management be avoided", indicating that they were slightly less reliable than the Belgian establishments.

72 Renwick.
Marble and Marble Working.
London 1909.

199

Fig 11.5 shows one of the more sophisticated machine tools of the time for facing blocks of marble.

Several important French companies were active at the turn of the century. L'Exploitation des Marbres de l'Ouest had a large factory at Quemont that boasted 840 saw blades, with works at Sable in Brittany that were even bigger. Other companies were Grand Marberie at Bigorre; Henri Vienne of Cousolre (see Fig 11.6); Devillers et Compagnie of Marpent; and finally M. Cantini in Marseilles. These companies may or may not have produced clock cases (I have found no record); their names are included for reference.

Marble Production in Belgium

Fig 11.6 Interior of the marble machine shop at Cousolre, France.

Belgium has large deposits of black marble, or more correctly calcareous marble. The stone was quarried in the territory from Roman times and one source notes that the Romans were quarrying there in the year 55 BC.

One peculiarity of the Continental system of marble working was that Belgium, a country with limited natural resources of decorative marble, had some of the largest marble working factories in Europe. To add to the riddle, these production units were not located next to any large mines or quarries but were often some distance away.

Fig 11.7 A Belgian marble workshop at La Bussiere.

The Belgian marble processing industry was founded around the time of the first French Revolution in 1789. A possible reason for this is that the area was fairly stable politically in contrast to France and central and southern Europe. The industry did not expand rapidly, but the Belgians slowly and surely learnt the trade that came to the fore in the revival of marble towards the end of the nineteenth century.

In 1908 one of the largest companies was La Société Anonyme de Merbes-le-Château. They had four factories in Belgium, one in Germany, another in France, and a depot in Italy. Most of the work for Great Britain is said

201

to have come from their factory in La Bussiere in the Soir et Sambe Valley, Fig 11.7. Another large supplier to Great Britain was De Jaiffe Frères at Mazy. In this instance the factory was situated fairly close to the area that still produces Belgian black marble.

At the end of the nineteenth century the Belgian marble industry commanded the respect of Europe and usually obtained the best of any contracts that were available. Although the principle factories were a relatively long way from the seaports, they still managed to compete in terms of price.

In its time, the industry was considered a shining example of achievement by a small country working with energy and perserverance to build one of the largest operations in Europe. Belgium herself did not possess particularly attractive stone, and imported marble from other areas of Europe. One company in the early 1900s had sales in England worth some £60,000 a year of which only two-thirds represented products using native Belgian stone. At the end of the nineteenth century the Belgians were providing most of the finished work for Great Britain.

Germany as well as France imposed heavy import duties on Belgian marble, which had a severe affect on the regions where the stone was particularly difficult to extract, such as Dinant. The British had at one time considered instituting tariffs but there were divided opinions about their value. Although it had little if any effect on the Belgian industry, the Americans also introduced import duties on marble which significantly boosted their domestic production.

Clock Case Production in Belgium

The working of marble was a specialised process with each factory tending to work solely on one type of product. One would specialise in clock cases and articles that could be produced by the hundred, another would concentrate on chimney-pieces or fireplaces as they are called today. Others would produce marble table tops by the thousand.

It is unfortunate that more information is not currently available about case makers; time and good fortune will hopefully improve this situation. I spent many months seeking positive information about case making in Belgium, but found nothing of substance, apart from the knowledge that a lot of the marble used in cases came from the Mazy area of Belgium. This side of the industry was small compared with the amount of marble that was processed for architectural work.

Most of the Belgian marbles are from the Carboniferous and Devonian Age. They are not particularly attractive and, apart from clock cases, were often used for the more utility grades of building work such as shop-fitting, table tops, and sanitary fittings.

The carboniferous formations were considered to be exceptionally sound, with Belgian Black being the most popular of all black marbles. This fine stone was used in the production of French marble clocks.

Golzinne or Mazy Black Marble

At the end of the nineteenth century the Belgian Black deposits were predominantly in the Golzinnes area, and stretched some eight miles from Villaret-Sous-Saint-Martin-Balatre through Falnuee, Mazy, Isnes, Rhisnes to Emines. The deposits varied between nine and twelve metres in depth with an incline of about 18°. The black marble was found in regular layers separated by worthless bituminous shale. The presence of the shale confirms previous comments about the marble being found near coal-producing areas. Marble deposits varied in depth from about 50 mm to about 1.3 metres. The deeper beds, yielding the best stone, were usually the lower formations.

In 1825 P.F. Cauchy[32] noted only three black marble quarries in the Mazy area. This number rapidly increased as the century went by. The depth and nature of the good formations required mining operations. These were first started by J. Etienne in 1859. Blocks of marble were loosened with explosives and then broken up using bars. Flooding was a constant problem. One manufacturer at the turn of the century had a mine some 76 metres (250 feet) deep with pumps working at the rate of 50 cubic metres (1,750 cu ft) an hour.

By 1983, just one working mine was left in Mazy: Etablissments Dejaiffe, undoubtedly succeeding the earlier company of Dejaiffe Frères. Their marble is marketed under the name of 'Noir Belge de Mazy' or 'Mazy Belgian Black'. This was the stone that an English importer had in stock when asked about Belgian Black marble, during my research for this book.   Although

32 Cauchy.
Memoire sur la Constitution Geoglogique de la Province de Namur, Brussels, 1825

specific references to clock cases are rare, J. Demaret[43] in 1886 noted that the marble from this area had particular uses, with clock cases being among them.

The Black Marbles of Dinant and Theux

The industry in Dinant dates from the Rennaisance period, its marbles being considered among the finest available, with funeral work being one of the oldest uses. The difficulty of extracting the material, coupled with the trade tariffs imposed by Germany and France, led to the demise of the industry. By 1935 the last of the Dinant workings had closed down.

In 1828, one writer noted that Theux marble was possibly one of the best available and was particularly admired by sculptors. Another, Davreux[39], wrote in 1883:

"This marble could be considered to be one of the finest available and reminds us of 'noir antique' or of 'de Lucullus'. It takes an extremely brilliant polish.....This marble is mainly used in Paris for making pedestals, vases and clocks etc.".

Noir antique and de Lucullus were particularly fine true marbles, and not calcareous marbles.

Further evidence of its use in clock making is provided by a commercial directory, L'Almanach de Commerce 1826, which contains a reference to Theux marble being used in the making of clocks, plinths, fireplaces, columns, tombs, staircases, and washstands.

Marble was extracted from Theux before the 16th century but completely stopped in the second half of the 18th. Work was resumed under the encouragement of the politician, L. F. Dethier, who managed to generate some enthusiasm for a new company that was formed in 1844, one year after his death in 1843.

Grades of Belgian Black

Renwick[72] noted that Belgian Black was sold in four qualities; best, second best, common, and inferior. Unfortunately he did not qualify his statement by giving examples of each grade. At first it might be assumed that the four grades related to the density and blackness of each sample, they actually refer to the amount of white streaking that is present in a piece of black marble.

In 1973 the Mazy marble was graded by Etablissments Dejaiffe into noir fin, demi-fin, ordinaire,

43 Demaret
Gisements, Exploitation Travail et Commerce des Marbres en Belgique et a la Entranger. Belgium. 1866.

39 Davreux
Essai sur la Constitution Geognostique de la Province de Liege. Brussels. 1833.

72 Renwick
Marble and Marble Working: A Handbook for Architects, Sculptors, Quarry Owners and Workers, and all engaged in the decorative Industries. London. 1909.

courant, et batiment. (First class, second class, ordinary, and building quality). In 1933 Paul Dumon[45] of the Belgian Geological Society gave a more precise definition for the grading of Belgian black marbles which, in translation, reads:

"Black marble may be classified into four grades from I for the best through to IV for the poorest. Grades I and II should have a fine grain, even texture and be uniformly black. There should not be any sign of white veins or spots. Grade I will be able to take a particularly brilliant polish. Grade III will be similar to II with the exception of the odd white veining and spots. Grade IV will be less black, has a coarser grain with the occasional brown or grey marking."

Grades of Marble used in Clock Cases

Although some cases are made of superb and dense black marble with a beautiful lustre, others are of very poor quality although definitely made of marble and not slate. Very good quality French marble clocks, such as that in Fig 3.1, were undoubtedly made of the finest stone that was available in Belgium, possibly the Theux variety.

The cheaper stones that did not take a good polish were probably used to make many of the cases that are waxed or lacquered. When this coating is stripped off, the marble often has a grey spotty appearance suggesting that the cases were of poor quality to start with.

Belgium was also famous for some beautiful red marbles that are found occasionally in either the decoration of French marble clocks or used to make the complete case. One of the best known was Rouge de Rance, which is classed as a decorative marble. At one time stone was quarried from Rance for the Palace of Louis XIV at Versailles. The quarry was closed, only to be opened again in 1900 to provide 120 monolithic columns for the construction of Antwerp Railway Station.

The Belgian marble industry supplied the bulk of quarrying machinery for continental Europe. The manufacturers were centred around Namur and Charleroi. The Belgians were often the first to try out a new technique. The wire saw was widely used in conjunction with the penetrating pulley or with shafts that had been sunk by revolving drills. These allowed

45 Dumon.
Quelques Observations sur le Marbre Noir de Mazy. Belgium. 1933.

complete marble columns to be cut in one drilling as shown in Fig 11.3. The only limitation to the size of these columns was the weight that could be raised by the lifting tackle.

Belgium was one of the first countries to mine for marble and continued to use the same methods until modern times. It seems that black marble was the main product.

Marble Production in the British Isles

Britain possesses a suprisingly wide variety of marbles. During the nineteenth century it could, with the exception of white marble, easily have satisfied domestic requirements. The working of marble dates a long way back and references were made in 1577 to marble quarries in Staffordshire. The quantity used in clock cases was, however, very small when compared with that from Belgium.

The finest black marble was said to have come from Ashford, Bakewell, and Derby. These three towns are all in Derbyshire. Welton in Staffordshire produced a coarse black marble, while the stone from Ribblesdale, Sedbury, and the western moorlands of Yorkshire, was considered to be durable. Port St. Mary in the Isle of Man produced a hard, long-lasting stone that took a good polish.

The British black marbles could not be quarried in large fault free blocks although the colour was usually considered to be good. The best quality occured in deposits of between 100 mm and 500 mm in depth. The black marble quarried in the area of Ashford in Derbyshire had an exceptionally close texture but lay in shallow beds and was even harder to work than Belgian Black. In addition, it was more expensive than the Belgian variety.

British Marble Clock Cases

In search of information, I went to Derbyshire County Library, which suggested the National Trust, who advised me of a paper written by Dr Trevor D. Ford of the Department of Geology at the University of Leicester. His paper, 'The Black Marble of Ashford-in-the-Water, Derbyshire', gives a detailed account of marble working in the area. He told me that marble from this district had been used for clock case making and that one of his relatives owned a clock with a case made of Ashford or Buxton marble. Shown in Fig 11.8, it is a beautiful example of its kind with delicate inlay work on the front of the case. The movement, striking on a bell, is probably French, dating from around the end of the 19th century.

Fig 11.8 A fine example of an English Derbyshire marble clock. Note the delicate inlay work that is typical of marble working in the Derbyshire area. Photo: Dr. T.D. Ford.

The inlay on the plinth is white shell, sliced up for the petals of the jasmine, and green marble for the leaves, specially imported as the green ornamental stone malachite was expensive. The inlay on the flanks depicts forget-me-nots, the specially prepared blue material being from Butterly ironworks near Derby. The leaves for this are Florentine green.

Black marble was quarried from the area surrounding Ashford-in-the-Water from the 16th century until production ceased in 1905. The heyday of the marble industry was during the Victorian period when demand led to the stone being mined underground. In 1748, Henry Watson established a marble mill in Ashford that produced a wide range of ornamental products for buildings, statuettes, vases and table tops.

Most of the marble was originally quarried from the Arrock Quarry and later mined underground. The rough blocks of marble were taken across the River Wye to be sawn to the required shape in a water powered mill before being ground and finally polished. The year 1835 saw the start of the inlay work such as that seen in Fig 11.8, coloured and ornamental stones being provided from neighbouring quarries.

Another source of interesting information is the Peak District Mine Historical Society who have a museum in this area.

Fig 11.9 An English marble quarry at the end of the nineteenth century. It was in Ashburton, Devonshire.

British Production Methods

The methods employed in Britain were usually crude and lagged far behind those of France and Belgium. Blasting was a popular technique that destabilised the stone, giving some British marble a reputation for unsoundness. The British quarry in Fig 11.9 is in stark contrast to the Belgian operation in Fig 11.10. The wire saw and other sophisticated techniques employed on the Continent do not appear to have been used at all in Great Britain.

British owners were usually unwilling to develop their quarries on the scale that existed in Belgium and France. They were equally unwilling to generate any stock of marble for future use. The quarries tended to use unsophisticated production methods that produced

Fig 11.10 The Petit Granit Quarry at Soignies, Belgium.

stone of ill-assorted size and quality. This, coupled with the fact that some of the best marble came from quarries that worked for only three months of the year, led architects, for example, to specify Belgian and French marbles in preference to locally-produced stone.

Another factor that held the industry back was the high cost of transporting materials by rail. At the beginning of this century the freight charges from Derby to London were higher than shipping the same stone from Belgium or Italy. The inefficent quarrying methods in Britain did not produce sized blocks, which meant that the buyers were paying to transport a lot of material that would only be wasted. Import duties were considered at one time as a means of securing more work for the British industry but met with divided opinions on their value.

The situation was summed up in these words in 1909 by W. G. Renwick, an English writer of the time:

"The manufacture of clock cases gives employment to a large number of hands in both France and Belgium. The work is specialised, the workmanship and finish of these being particularly good. Clock cases are produced at such exceptionally low prices as to make competition impossible in this country."

Irish Marble

Deposits of black marble were sited at Auglihan in Galway and Menlough, Ireland, where large blocks could be obtained. The marble from Kilkenny is said to have taken a good polish but after a time the blackness faded to reveal small white fossils on the surface. Other quarries were at Crayleath in County Down, and Churchtown and Doneraile in County Cork. Carlow, near Tralee, and the islands of Kenmare River also produced black marble.

209

ELLIS & OWEN,

ABERYSTWITH PLAIN & ENAMELLED SLATE & MARBLE WORKS,

ABERYSTWITH,

MANUFACTURERS OF

ENAMELLED & PLAIN SLATE CHIMNEYPIECES,

BATHS, CISTERNS, MILK COOLERS,

HEADSTONES,

URINALS, MANGERS, CATTLE TROUGHS, DAIRY, LARDER, AND WINE CELLAR
SHELVES, WINDOW SILLS, DOOR STEPS, HEARTH STONES, FLOORING,
SKIRTINGS, AND EVERY DESCRIPTION OF

Plain and **Enamelled** **Slate** **Work.**

Slate Enamelled in imitation of the most costly Marbles, at a comparatively low
price, for Chimneypieces, Hall Tables, Sideboards, Consoles, Cheffoniers, &c. Circular
Table Tops, Finger Plates for Doors, and Paper Weights with Views, Escutcheons,
Chess, Flowers, Ornamental and Marbled.

CRESTS, &c. ENAMELLED ON KEYSTONES OF CHIMNEYPIECES.

AN ARTIST SENT TO TAKE VIEWS WHEN REQUIRED.

DESIGNS OF EVERY DESCRIPTION EXECUTED TO ORDER.

Roofing Slates of all sizes always in Stock.

☞ *The First Prize for Slate Carving was awarded to this Firm both at the Chester Exhibition*
(1866) and the Carmarthen Exhibition (1867) of the National Eisteddfod of Wales.

HENRY P. HAWKINS,

PROFESSIONAL HAIR CUTTER,

AND COURT HAIR DRESSER,

ARTIST IN HAIR,

23, PIER STREET, ABERYSTWITH.

M. VAUGHAN REES,

LAPIDARY & WORKING JEWELLER,

14, Marine Terrace, and 1, King Street, ABERYSTWITH,

Begs to return sincere thanks to her friends and the public, and also to the visitors, for the kind patronage she
has received during the last 41 years, and respectfully solicits a continuance of their kind support.
A most extensive Stock of Aberystwith Mounted Pebbles. Jewellery, &c. Carefully Repaired.
Variegated Specimens of Lapidary Jewellery may be inspected at her Establishment.
The utmost value given for old Gold and Silver.

Fig 11.11 An advertisement in Worrall's Trade Directory of North
Wales (1874). Ellis and Owen were capable of enamelling slate in
imitation of marble.

Welsh Slate Cases

I contacted various institutions throughout Wales, after a dealer at Covent Garden had shown me a case that he assured me was made of Welsh slate. This was before I invented the acid test described in Chapter Ten. Unfortunately I never managed to find a definite reference to clock case production although Inigo Jones, Slateworkers of Caernarvon, Gwynedd, Wales, had a case made of slate in their office. Sadly it was not fitted with a movement, giving little clue to its background. It is possible that the Welsh supplied polished and unpolished slate pieces to case makers in London and elsewhere in Britain. It is also a possibility that the local makers of chimney pieces produced clock cases as well.

Fig 11.11 shows an advertisement from the 1874 Worrall's Trade Directory of North Wales, in which Ellis and Owen refer to their ability to enamel slate 'in imitation of the most costly Marbles at a comparatively low price' and the fact that they won awards in 1866 and 1867 for the quality of their slate carving.

There is little doubt that only in London was there any significant marble case making.

XII

Restoring Marble Cases

Marble is normally a difficult sone to work, and Belgium black is particular is hard to work and finish. Much of the shaping of mouldings and cylinders for the cases would have been performed on a machine, although carving was usually undertaken by hand. Machines existed in the early 1900s that could do repetitive carving work, but the bulk of carving was still left to the hands of a skilled mason.

If the work was being undertaken by hand, the approximate shape required would be formed with a saw and chisel. Circular objects, such as the drum shapes in some marble clocks, were finished on a lathe. The finished product would still be rough to the touch and required further smoothing with abrasives.

Fig 12.1 A marble shaping machine, circa 1903. The cutting action was provided by carborundum stones revolving at high speed.

Most simple clock cases were made by machine. There were two distinct methods for producing marble mouldings; one machine employed a scraping action, while a later version used carborundum stones to grind the marble away.

The scraping machine was heavy and the machine tools were the shape of the moulding. The tools were set side by side in a massive box that ran the width of the machine as seen in Fig 11.5. The carborundum machines came into use at the start of the twentieth century. They were relatively light in construction and worked at a higher speed than the scraping machines. The device in Fig 12.1 could shape a one metre length of cut marble into a moulding suitable for use on a clock case in about five minutes. The finished product was suffcently smooth to save the polisher a considerable amount of preparation time.

213

Polishing a Marble Case

When I started research for this chapter, I hoped to discover a perfect method for restoring marble cases, but failed, unfortunately. Each member of the marble polishing community uses his own particular system and there is no common process that is either simple or straightforward. If you have a valuable case, I would suggest that you either leave it well alone or contact a marble mason.

All the information I was able to amass is given here and I hope the reader will be able to evolve his own system. I recommend strongly that the back of a case be used for practice before attempting work on the more visible sections.

Over the years I have tried various methods in the hope of producing a fine sheen on the black cases of marble clocks, using such diverse products as boot polish, French polish, lamp black, beeswax, linseed oil, and shoe dye. With the exception of beeswax, these substances were generally of little value. The process I use now requires wire wool, slate blacking, turpentine, and wax, and appears to be the safest and most effective. This finish is also used by a fellow enthusiast whose knowledge and experience I have come to value.

Polishing New Marble

There are at least two nineteenth century accounts of marble polishing, one in 'Spon's Workshop Receipts'[78] (1895) and the other in an American book, 'The Marble Workers' Manual' (1886) by Mary L. Booth. The accounts are identical and it is assumed that they both used the same source text. Booth usually translated European works to provide badly-needed literature for the nineteenth century American craftsman.

To quote from Booth, complete polishing includes five distinct operations:

"1. Grinding, which consists in smoothing the roughness left by the burin [chisel]. This is done by rubbing the marble with a piece of moist sandstone; for mouldings, either wooden or iron mullars [shaped rubbers] are used, crushed and wet sandstone, or sand, more or less fine according to the degree of polish required, being thrown under them.

78 Spon. *Workshop Receipts: For the use of Manufacturers, Mechanics and Scientific Amateurs'. Spon, London. 1895.*

214

2. The second process is continued rubbing with pieces of unglazed faience [pottery], without enamel, which have been baked but once, also wet. If a brilliant polish is desired, Gothland stone instead of faience is used, and potter's clay or fuller's earth, a sort of clay mixed with fine sand, is placed beneath the muller.

Perfection of polish depends almost entirely on the care bestowed upon these two operations which should be performed with a regular movement, requiring much patience.

When the marble has received this first polish, the flaws, cavities, and soft spots are sought out, and filled with mastic of a suitable colour. This mastic is usually composed of a mixture of yellow wax, resin and Burgundy pitch, mixed with a little sulphur and plaster passed through a fine sieve, which gives it the consistency of a thick paste. To colour this paste to a tone analogous to the ground, tints or natural cement of the material upon which it is placed, lamp black and rouge, with a little of the prevailing colour of the material are added.

3. The third operation of polishing consists in rubbing it [the marble] again with a hard pumice stone, under which water is constantly poured, unmixed with sand or other mordant [abrasive].

4. For the fourth process, which marble workers call softening the ground, lead filings are mixed with the emery mud produced by the polishing of mirrors or the working of precious stones, and the marble is rubbed with a compact linen cushion well saturated with this mixture; the English rouge is also used for the first polish. For some outside works, and for hearths, paving tiles, etc, marble workers confine themselves to this finish.

5. Finally, in order to give a perfect brilliancy to the polish, the gloss is applied. This is done by first washing well the prepared surfaces, and leaving them until perfectly dry; then take again, a linen cushion, moistened only with water, and a little powder of calcined tin of the first quality. After rubbing with this for some time, take another cushion of dry rags, rub it lightly, taking care to brush away any foreign substance which may crease [scratch] the marble, and a perfect polish will be obtained."

215

Booth continues by mentioning a quick method of polishing that impairs the durability of the polish:

"It is necessary to observe that, in order to gain time and facilitate labour, many marble workers mix alum in the water which they use. This mordant penetrates the pores of the marble, and really gives it a speedier polish. This, however, is a fictitious polish, which spots very easily, and which is soon tarnished and destroyed by dampness."

Modern Polishing Methods

The modern version of the traditional polishing process is to grind the surface with an abrasive, starting with a coarse grade of carborundum stone and ending up with fine one. The final gloss is obtained by polishing with calcinated tin, which is sometimes called 'potate of tin' or 'putty of tin'. Some authorities suggest a fine emery powder, called 'flour emery', as an alternative. New marbles would have been finished by one of these.

Restoring Old Marble

The method described above is fine for polishing new marble, but is not particularly suitable for restoring an old case made of calcareous marble, which is not a true marble. For most purposes, differences between the stones can be ignored, but not when restoring the polish of an old case. Calcareous stones, of which Belgium black marble is one, do not appear to take to the polishing process that is applied to normal true marbles. I have experimented in some depth with various traditional techniques including a modern version of Mary Booth's. Some cases does not respond to being polished with traditional materials. Fortunately there are alternatives.

Before attempting any work on the case, take an eyeglass and look at it carefully. At this stage, try to answer three questions:-

1. Has the case been lacquered with a paint or similar material?

2. Is the surface finish a wax polish that looks like a French polish?

3. Is the case completely free of any surface finish?

216

It is important to consider the above questions before proceeding with any work on the case. In the early days I cheerfully stripped the coating off a case, to find that the stone underneath was of poor quality. I never really managed to get rid of the whitish appearance of this stone and would have been better off with the original coating. Take this then as a warning: If the case has been coated in some way, leave it alone unless you have good reason to strip it.

Restoring a Case with Wax and Slate Blacking

The simplest way to restore a case is to wax it, using either beeswax or a proprietary black marble wax coupled with a base coat of slate blacking. The latter is available from Meadows and Passmore Limited, of Crowborough, England. Wax has been used successfully by enthusiasts for many years. It is also probable that wax was used by the ancient Greeks. As Mary L. Booth put it, "I have heard of a kind of varnish made of white wax highly extolled as a preservative; this is laid on by means of heat and is afterwards rubbed with a cushion. It is said that this varnish was used by the ancients [Greeks] and that the preservation of their chefs d'oeuvre [masterpieces] may be attributed to this".

Spon mentioned a wax finish similar to Booth's, "Heat a half gallon of water, in which dissolve one and a half pounds of potash, add one pound of virgin wax, boiling the whole for half an hour, then allow it to cool, when the wax will float on the surface. Put the wax into a mortar and triturate [crush] it until it form a soft paste, which laid on marble, and rubbed, when dry, with a woollen rag, gives a good polish".

In his book 'La Pendule Française', Tardy[80] says that a substance called 'stearin' was used to coat the black cases. A chemical dictionary describes stearin as "…. a component of the solid fats. When pure it is a colourless, tasteless substance, soluble in ether but insoluble in water".

On the same page, the dictionary notes that stearic acid and its derivatives are used as lubricants and protective coatings. It is possible that stearin is the very thin hard finish seen on some cases.

It is fortunate for the modern worker that ready made black waxes (Fig 12.2) are obtainable from the suppliers

80 Tardy.
La Pendule Française.
Paris 1970.

217

of marble working materials. These smell similar to beeswax, are easily applied, and if not applied too heavily should produce a fine waterproof sheen.

It is possible to make beeswax polish, but it will be untinted. Beeswax may be bought in small blocks from good hardware shops and artists material dealers. Some genuine turpentine will also be needed. A similar liquid sold as white spirit or turps substitute, will *not* do, only genuine turpentine. The wax is prepared by putting 250 grammes in about 150 ml of pure turpentine. Heat it slowly in a metal container placed in hot water and wait until it has completely dissolved. Additional turpentine may be added until the liquid is fairly sloppy.

Fig 12.2 *Old and modern marble waxes. On the left is a contemporary black marble wax and on the right a block of beeswax, which would be diluted with turpentine in the bottle.*

Waxing and Blacking a Marble Case

The surface of the marble must first be cleaned with a clean lint-free cloth that has been dampened with pure turpentine and formed into a pad. Gently rub the case with the pad, taking care to get into the corners of the mouldings; turn the pad occasionally to reveal a clean piece of pad. When satisfied that the case is clean of grime and dirt, leave it for a while to dry. Fine No.0000 wire wool, dipped in turpentine, may also be used to help clean off the accumulated dirt but take care; if possible, use the cloth and turpentine first.

Before applying any wax a base coat of slate blacking should usually be applied. This may not be necessary on some of the top grade marbles but will make a dramatic improvement to the poorer grades. Apply the blacking using the instructions that are given here for waxing. Leave the blacking to dry for as long as possible before applying any wax.

The most important point to remember when applying wax is to use only a little, and work it well over the surface of the stone. There is a temptation to apply a large quantity in the hope that a better polish will be obtained. Do not bother trying; the results can be unpredictable.

Take a fresh pad and apply a little wax to the flat surfaces of the case, using a circular rubbing motion followed by long strokes. Allow the wax to dry completely before obtaining the final polish by buffing with a clean soft cloth. If a hard, or patterned, cloth is used there will be a tendency for the surface to be marked or scratched. The application of a little wax may be made several times until a deep enough sheen or 'body' is obtained on the marble.

After polishing, the surface will still be fairly soft and it is advisable to leave the case untouched for a period of several weeks, so that the surface coating hardens, before attempting any further work on it.

Maintenance of a Waxed Case

In normal circumstances, a waxed case should only need a dusting and the occasional rub with a soft cloth. Additional coats of wax will not be required. Modern spray polishes should be avoided at all costs; they usually contain silicone, which will impair the finish of the marble.

Many flat-topped cases have white rings on them, left by wet drinking glasses. The only effective way to remove this sort of stain is to grind the surface with a fine abrasive until a new surface appears. Use the finest wet-and-dry paper available, after taking away the sharpness of new paper by rubbing the abrasive surfaces together. On no account should a coarse paper be used,

219

otherwise the marble will be ruined.

Ideally the paper should be used with a rubbing block and be kept wet at all times. A gentle circular rubbing motion will eventually abrade the surface sufficently to reveal a new one. This may be sufficently smooth to take a polish with wax, or may require further finishing with wet and dry paper until it is completely smooth.

Slate cases were smoothed with abrasives before having a final polish with powdered emery. Then they were lacquered. Nowadays the recommendations made for waxing a marble case can be followed. The lacquer should only be stripped if it is deemed to be genuinely necessary.

Repairing and Restoring Bezels

There are two bezels on a French marble clock, one at the back and the dial bezel at the front. The normal problems associated with bezels are dealing with broken hinge pins and glasses, and the need for cleaning.

Occasionally the hinge on a bezel will have broken, leaving the steel hinge pin in the hinge itself. The pin will often be rusted in position and considerable care will have to be taken to ensure that the brasswork is not damaged. Try soaking the hinge in releasing fluid and then attempting to tap the pin out, remembering that the pins are tapered with the thicker end in the upper part of the hinge. A replacement pin is made out of a length of steel that is tapered with a file until it is a snug fit in the hinge plates. Fine fitting may be achieved by using a broach to open the hinge plate holes.

Personally I do not clean rear bezels. They are normally painted black and a rub with a black shoe brush normally polishes them without affecting their aged appearance. The front bezel will invariably need to be polished and care should be taken not to turn a softly aged piece of brass into a piece of gleaming metalwork; if the brass is still in good condition, leave it alone.

If the front bezel is to cleaned, remove any lacquer originally applied to stop the brasswork tarnishing. This work is sometimes easier if the glass is removed, as explained below. The lacquer is easily removed with a water soluble paint stripper, such as Nitromors. After washing in water and drying, the brass can either be polished with a conventional cleaner, or dipped in the cleaning solution mentioned in Chapter Seven. When the bezel is clean, lacquer it with a clear lacquer, using a fine brush. This work should be performed in a warm and dust free place.

Fig 12.3 Removing a bezel glass. The brass bezel expands quicker than the glass, allowing the glass to be gently pressed out.

Removing a Bezel Glass

Removing a bezel glass is one of those jobs that is easy when you know the correct way. The glass may be broken or you may wish to remove it for cleaning; the procedure

221

is the same for both jobs. Place the bezel near the flame of a stove ring and gently heat it, using the method shown in Fig 12.3. Alternatively a blow torch may be used.

Take care not to get the bezel too close to the heat source, or the glass may crack. After a while you should be able to pop the glass out of its mount because the brass expands more than glass. A broken glass may be measured and a replacement obtained from a clockmakers' material dealer. The glass is replaced by using the same procedure, but this time by pushing a cold glass gently into a warm surround. If the glass is still loose, it may be held with a clear adhesive.

Collectors' Clocks

Thirty-two clocks are illustrated here to give an idea of the diverse range of shapes and sizes that are still available to the enthusiast. I have not attempted to give any dates to the pieces except where I am sure of my facts.

*Fig 1. **Private collection.** A fine marble clock that is fitted with the unusual Brocot two-wheeled visible escapement. The case has an attractive shape, enhanced by the well made mitre joints in the base. The striking is by rack and bell with the rack being raised by a lever on the end of the hammer. The movement was probably made by Brocot and retailed by Leroy. Movement number 126, signed 'Leroy et Fils' and numbered 3779, probably by Leroy. The bezel is stamped 126 and has the Brocot five-pointed star stamped on to it. Brocot regulator without click. Solid zinc pendulum numbered 126.*
Plate diameter 10 cm. Height 38 cm (15 in).

*Fig 2. **Photograph: Sotheby's, London.** A splendid example of a French marble clock in the Egyptian style. The statuette of Cleopatra is in bronze. It has a square plated movement striking on a gong and numbered 72. A Brocot escapement is fitted internally.*
Height 64 cm (25 in). Circa 1880.

*Fig 3. **Private collection.** This clock shows its quality in the amount of work that was required to make it. The key points to note are the bevelled dial glass, the carefully made columns on the sides, and the delicate incised work on the case. A few small pieces of coloured stone complete the decorations of a fine specimen. The movement is stamped 'S. Marti et Cie. Médaille de Bronze' with the number 1915. A Thieble pendulum is fitted. The bell has a repair mark of 1886, although this would be unreliable for dating purposes as bells are sometimes moved from one clock to another during repair. Plate diameter 9.6 cm. Height 33 cm (13 in).*

Fig 4.
Private collection.
A fine clock with outstanding features that would always make it a popular piece. The drum surrounding the movement is in green marble that also appears in the sides and base. The Ellicott pendulum is an attractive accessory, as is the Brocot visible escapement. A black wooden door was removed from the rear of the base before the photograph was taken. The rear movement bezel is stamped 'C.L.F.' The clock is numbered 1330, with rack striking on a bell. Plate diameter 9.6 cm. Height 43 cm (17 in).

225

*Fig 5. **Private collection.** A plain black marble case that is reminiscent of a regulator type of clock. The mercury pendulum is an attractive feature, as is the visible escapement. Note that the bars of the latter are straight. Some authorities take note of whether these are straight or curved. The movement has a spring support for the hammer and may therefore be dated before 1870. Locking plate striking with a bell. The plate is numbered 5249. Plate diameter 11 cm. Height 46 cm (18 in).*

*Fig 6. **Private collection.** A large American Ansonia clock fitted with a visible escapement. The movement from this clock is shown in Fig 3.11. The overall design is crude and has a bizarre mixture of styles within it. Greek columns are mixed with other motifs, creating a clumsy piece that lacks the elegant simplicity of its French counterparts. American Ansonia movement striking on a gong. Note the low position of the winding holes that are indicative of an American movement. Height 41 cm (16 in).*

*Fig 7. **Private collection.** A fine clock that is a perfect specimen of its genre. This type of clock will always command a high price and would delight many enthusiasts. The perpetual calendar below the main dial shows the day, date, month and the lunar phases. Unfortunately the photograph is unable to do justice to the beautifully enamelled dial that has the sky and clouds depicted in various hues of blue. The movement is numbered 7569 and stamped with a star with 'UF' in its middle.*
Plate diameter 11 cm.   Height 43 cm (17 in).

Fig 8. **Private collection.** A classic example of a French marble clock with an English company name on the dial. The movement is stamped 'J.B.D.', the initials of the Frenchman Delettrez. The dial is marked 'Manoah Rhodes & Sons, Bradford'. The movement is of particularly good quality and has rack striking on a bell. The case is fairly plain and enhanced with red marble mouldings. Movement details: Stamped 'J.B.D. No.27210' and fitted with a large, heavy pendulum that is solid in construction.
Plate diameter 8.6 cm.
Height 25 cm (10 in).

Fig 9. *Private collection.* A good case with a green marble trim. Although the case is fairly plain the mouldings around the base indicate that it was a relatively expensive design to produce. This shape and style is seen in many variations of size and decoration. The dial is marked 'J.R.W. Chaplin, London Bridge' while the base of the case is indistinctly marked 'Mignier Marble'. It is fairly unusual to have any references to marble marked on the case.
The locking plate movement strikes on a bell. Plate markings: 'Japy Frères & Cie. Exposition 1855. (Grande Med. d'Honneur)', also stamped 'D'Alreville et Chaemhoy Mson.' Movement numbered 297.
Plate diameter 8.2 cm.
Height 23 cm (9 in).

*Fig 10a, 10b, and 10c. **Private collection.** (10a, bottom right): A tiny example of a plainer marble case. The case is about 20 cm high and contains a Japy locking plate movement that strikes on a bell and has a Thieble pendulum. Plate details: Number 396 marked 'Japy Fils Expos. 1855, Médailles 1844, 1849'. Plate diameter 81.5 cm. Height 20 cm (8 in). Circa 1860.*

(10b, bottom left): Another tiny clock that has some fine incised work on its front. The Japy rack striking movement strikes on a bell. Plate detail: Number 18243 stamped 'Japy Frères - Legion d'Honneur'. Plate diameter 8.2 cm. Height 62 cm (7.5 in). Circa 1885.

(10c middle rear): A large and heavy clock with a fine Marti rack striking movement. It has a Brocot visible escapement and Brocot pendulum. The movement is numbered 226 and stamped 'S Marti & Cie. Médailles de Bronze'. Plate diameter 9.6 cm. Height 43 cm (17 in). Circa 1875.

Fig 11. **Private collection.** *A magnificent clock that has been carefully restored by its owner. The case is inset with lapis lazuli and the dial is of a fine blue colour with gilded numerals and hands. The inner brass dial is ornately worked to complement the plainess of the outer dial. This fine example has an unsigned movement numbered 4494, complete with its original elliptical pendulum, and rack striking on a bell. The rear bezel has a triangular stamp with the letter 'J' in it.*
Plate diameter 10 cm. Height 47 cm (18.5 in). Circa 1870.

Fig 12. **Private Collection.** *An unusual case made of black and green marble standing some 18 inches high houses the movement of this clock, which has an Ellicott pendulum and is marked 'S Marti & Cie, Médaille de Bronze' and numbered 3599.*
Locking plate striking with a bell.
Plate diameter 96 cm. Height 46 cm (18 in).
Circa 1875.

*Fig 13. **Private collection.** A magnificent clock with a dial that is almost identical to Fig 12 although the movement has square plates, a relatively unusual feature for French marble clocks. The case is 51 cm high and shows the use of Greek columns with scenes above and below the dial, presumably of Greek origin. This clock has been dated as circa 1885 and marks the beginning of the so-called 'Greek column' period. By the early 1900s, columns appeared in the majority of clocks. Once again the simple case design could have been made entirely by machine. The square plates are numbered 2919 and are unsigned.*
Rack striking on a gong that is stamped 'PS'. Thieble pendulum.
Plate width 8.3 cm. Height 51 cm (20 in). Circa 1885.

Fig 14. **Private collection.** *A plain yet dignified case in black marble trimmed with green. The grooving on the front reduces the overall weight of the design; a thick and bevelled dial glass creates an impression of solidity. Unsigned movement numbered 283 and 180. Elliptical pendulum. Rear bezel stamped with a 'J' inside a triangle (see caption of Fig 17). Plate diameter 9.6 cm.*
Height 36 cm (14 in). Circa 1880.

Fig 15. **Private collection.** *This clock is a classic example of one that appears to be of high quality while it has been made using simple production methods. In Chapter Ten it was said that a simplicity of design led to a reduction in manufacturing costs. A close look at this example will show that it is made out of simple slabs of marble. With the exception of the incised work, the whole case could have been made by machine and would not have required any hand carving by a skilled craftsman. The incised work would have required some skill although this was sometimes etched with acid. The dial is only partly made of enamel and certainly dates from either the end of the nineteenth century or possibly the beginning of the twentieth. These dials were, in my opinion, cheaper to make than a full dial and are often seen on poorer quality clocks. The movement is marked 'F. Marti-Médaille de Bronze' and numbered 878. The front plate is attached to the pillars by screws, another indication of the poorer quality of the overall clock. Rack striking on a gong. Plate diameter 8.2 cm. Height 30 cm (12in). Circa 1900.*

*Fig 16. **Private collection.** Standing a foot high, this clock is a good example of its type. The volutes or grooves in the upper surfaces of the case reduce the overall heaviness of the design. They are reminiscent of the volutes seen in Greek columns and are frequently found on this style of case. Brown marble has been used extensively to break up the darkness of the black marble. A close look at the base of the case will show cracks, evidence of being dropped at sometime. Such damage is not unusual and is a point that buyers should be aware of. The clock has a large movement bearing a stamp 'Vincenti & Cie, Médaille d'Argent 1855'. It is numbered 512 and has rack striking on a bell. Plate diameter 11 cm. Height 30 cm (12 in).*

*Fig 17. **Private collection.** A magnificent example of the marble mason's craft. Most of the upper part of this case would have been hand carved in a fine black marble of sculpting quality. The front elevations are inset with red marble giving the overall design a richness and elegance that is characteristic of good French marble clocks. Numbered 2399 and stamped 'S. Marti & Cie. Médaille de Bronze', the rack movement strikes on a bell. The bezels are stamped with a 'J' in a triangle. Plate diameter 9.6 cm. Height 33 cm (13 in).*

Fig 18. **Private collection.** *Another magnificent clock with a carefully sculpted upper case that is typical of the drum style. Note the absence of a Brocot visible escapement, something in my opinion that adds to its charm rather than detracts from it. The visible escapement can be an interesting feature but is also distracting in a simple and elegant case design. Locking plate movement striking on a bell and stamped 'Japy Frères & Cie. Gde Médaille d'Honneur'. Brocot escapement, solid pendulum that carries a scratched repair date of 13th May 1879. Height 28 cm (11 in). Circa 1870.*

*Fig 19. **Private collection.** The large case, 16 inches high, is not unlike some American clocks (see Fig 6), although it is of French origin. The blue dial is similar to those in Plates 14 and 16. the dial is marked 'Howell James & Co. Paris'. Jones would have been the English importer, who added respectability with the Paris address. Other English companies had Parisian marks on their clocks although it is doubtful whether they had little more than an agent in the City of Paris. Number 591, stamped 'S. Marti & Cie. Médaille de Bronze'. Brocot escapement with solid pendulum. Plate diameter 10 cm. Height 41 cm (16 in).*

236

Fig 20. **Photograph: Sotheby's, London.** *This superb piece is made of black Belgian marble and has a top quality French movement. The ornamentation is in bronze and the black case has been heavily inlaid with brown coloured marble. The enamel dials are set within an oval gilt surround, engraved and divided by porcelain plaques painted with cherubs against a pink ground. The movement by J.B. Delettrez is numbered 34569, with a visible Brocot escapement and gong striking. A lever connects the strike train to the calendar movement that indicates the date, day, phase of the Moon, and the equation of time. Height 58 cm (23 in).*

Fig 21a, 21b, and 21c. **Clock courtesy of Mr L. Glendenning. Photograph: Sotheby's of London.** *(a): An unusual and interesting year clock. The single winding hole indicates that it is a timepiece and does not strike. The plaque on the front of the case reads: 'Presented to Capt. Walter Paton commanding the steamship Great Eastern by the passengers on the voyage from New York to Liverpool, August 4th. 1862.' The dial is marked 'Regulator going one year without winding.'*

(21b and c): The movement is numbered 2487 and was made by A. Brocot. The rear plate bears the familiar Brocot star. Four ganged going barrels allow the movement to run for a year without winding. Height: 46 cm (18 in).

Fig 22. **Clock by courtesy of Mr P. Gregory.** *An attractive clock with a domed top piece. The four pillars are turned from a light green stone that could be onyx. Other parts of the case are similar to other clocks in this chapter, especially the incised work on the base. The movement has a medallion inscribed 'Medaille d'Argent' and the striking is rack controlled and sounds on a gong. Height 25 cm (10 in).*

Fig 23. **Private collection.** *A small clock built from sections of marble that would have been made by machine. The rack striking movement, numbered 4095, sounds on a gong. A platform escapement is fitted in place of the usual pendulum, indicating that the clock dates from a fairly late period. The English words 'Made In France' are the only other markings on the backplate. Height 25 cm (10 in). Circa 1910.*

239

Fig 24. Photograph: Sotheby's, London.
An example of what are commonly known as mystery
clocks. It is made out of black marble and metal, and
while not strictly a French marble clock, it has been
included for interest. The pendulum is given impulse by
the figure twisting slightly, being operated by a set of
rods that go down into the base of the case.
The dial is identical to others shown in this Chapter.
Height approximately 46 cm (18 in).

Fig 25. Clock courtesy of Mr R. Fortey.
This small clock is a splendid example of
its kind. The case is beautifully
proportioned and enhanced by the use of
coloured marble and squared corners.
A thick, bevelled dial glass helps to show
the plain white enamelled dial.
The dial is marked 'Hry. Marc, Paris'.
Stamped 'Japy Frères & Cie. Grande
Médaille d'Or'. The rack movement
strikes on a bell. Height 23 cm (9 in).

*Fig 26. **Photograph:
Sotheby's, London.** A white
marble clock fitted with a
perpetual calendar and
barometer. Standing some 18
inches high, this clock has
some Egyptian influence in its
design and was retailed in
Ireland by Lee & Son, Belfast.
It is numbered 28605 and
signed by J.B. Delettrez. An
Ellicott pendulum is visible
through a glass panel and the
calendar movement is driven
by the strike train. Note the
centre seconds hand.
Height 46 cm (18 in).*

*Fig 27. **Clock by courtesy of The Clock
Shop, Weybridge, England.** The black
marble case is inlaid with brown marble
and decorated with nautical motifs. The
gilt numerals stand on a black marble dial
and are complemented by gilt hands. The
movement is numbered 22689 and signed
'C. de Touche 228. Rue St Martin 230.
Gold Medal 1827'. A Brocot suspension is
employed and the movement strikes on a
bell. The gold medal of 1827 does not
imply that the clock dates from this period.
The Brocot suspension was not widely used
until after the 1850s. Height 25 cm (10 in).*

Fig 28. **Clock by courtesy of The Clock Shop, Weybridge, England.** One could be forgiven for thinking that this clock, viewed from the front, was yet another 'small and attractive French marble clock'. It is however a fine example of many clocks that had good French movements fitted into an ebonised wooden case. The movement is marked 'Vincenti & Cie' and is fitted with a Brocot suspension. It is numbered 1731 and stamped 'Vincenti & Cie. Médaille d'Argent 1855'. Height 18 cm (7 in).

Facing Page

Fig 29. **Private collection.** The ultimate French marble clock with a bronze figure. The marble is pure black and very fine, undoubtedly of the statuary quality mentioned in Chapter 5. Unfortunately the photograph tends to give a false perspective. In truth, the overall balance of the piece is magnificent and the large white dial blends well with the figure. Note the absence of a visible escapement that would tend to detract from the overall elegance of the piece. It is numbered 846 and stamped 'Vincenti & Cie. Médaille d'Argent 1855'. Another stamp gives the initials 'C.M.F'. The Brocot escapement is controlled by a solid pendulum. The original mainsprings are dated 1864. Plates 11 cm. Height 58 cm (23 in).

242

MAKERS and their MARKS

This collection of makers marks and names has been drawn from a wide variety of sources. Much information has come from my own research in France and Great Britain, but without the following reference works the list would have been much smaller: Carriage Clocks (Allix and Bonnert); Watch and Clockmakers Handbook (Booth); Clock and Watch Trademark Index (Kochman); Watch and Clockmakers of the World Volume 2 (Loomes); The International Dictionary of Clocks (Smith); Dictionnaire des Horlogers Français (Tardy); American Clocks for the Collector (Tyler). The Bibliography gives details of these books.

The list that follows gives names of people who have been associated with marble clocks. Some readers may question the inclusion of English, American and German makers. The Treaty of Frankfurt in 1870 allowed Germany to export unassembled movements into France free of duty and many collectors will come across marble clocks that have these poor quality movements fitted. To the untrained eye the clocks are often very similar to the the French originals. English and American makers have been included because it is difficult to hunt for French marble clocks without coming across interesting examples of marble clocks from these countries.

Dates of Exhibitions

It is fortunate that there were many national exhibitions in France during the nineteenth century where horologists, among others, were encouraged to exhibit their work. Many makers who received an award marked their movements with a medallion type of mark. For example, "JAPY FRERES Grande Médaille d'Honneur 1855" is often seen stamped on the back of a movement, indicating that Japy won an award at the Paris Exposition Universelle in 1855.

The original French exhibitions, the Expositions des Produits de l'Industrie Nationale, were held from 1798 onwards to celebrate the founding of the first French Republic. Few horoligists exhibited until after 1855. The Nationale type of exhibition eventually became Expositions Universelles where large numbers of clock and watch manufacturers exhibited their many products.

Abbreviations used in the reference section:

c (circa) = about;
Exp. – Exposition;
Grd. = Grande;
q.v. (quod vide) = which see.

245

The Expositions des Produits de l'Industrie Nationale were held in, or near Paris in: 1798, 1801, 1802, 1806, 1819, 1823, 1827, 1834, 1844, and 1849.

Expositions Universelles were also held in, or near Paris. The dates were: 1855, 1878, 1889, and 1900.

Similar exhibitions were held in London in 1851, 1854, and 1862. The Great Exhibition of 1851 is said to have encouraged the French to start their Expositions Universelles.

The awards given at the French exhibitions were placed in the following order of merit:-

Decoration de la Legion d'honneur (the highest).

Grande Médaille d'honneur (en Or) [gold].

Médailles de Première Class (Argent) [silver].

Médailles de Deuxième Classe [bronze].

Mentions honourables.

For more information on this subject see Allix and Bonnert, page 65.

ANSONIA. New York, 1850-1929.

An American clock manufacturer who produced a variety of 'marble' clocks. The back plate of the movements were usually marked ANSONIA CLOCK Co. U.S.A. NEW YORK. Dating is often aided by the stamping of the patent date on the back plate.

1850 - Ansonia Clock Company.
1869 - Ansonia Brass and Copper Company.
1877 - Ansonia Clock Company.

The company was dissolved in 1929 and the machinery sold to Russia. Fig 1.

Fig 1

ARNOLD, J.R. London, 84 The Strand, 1783-1840.

Maker of marble clocks with English movements. Arnold died in 1843 and was succeeded by Chas FRODSHAM (q.v.).

AUGUSTIN CROUTTE et Cie.
See CROUTTE, Augustin.

BARRAUD and LUND. London, Cornhill, 1838-1929.

Successor to family company of Barraud and Sons. Known to have produced a marble clock with an English movement c 1868.

BASCHET. Paris.

Received honourable mention for mantle clocks at the Paris Exhibition of 1855.

BAVEAUX Frères, Alfred and Louis. Saint-Nicolas-d'Aliermont.

Honourable mention, Paris Exhibition, 1878.

Silver Medal, Paris 1889.

Fig 2

BAYARD. Saint-Nicolas-d'Aliermont.

Originally founded by Albert VILLON (q.v.), eventually became Duverdy & Bloquel. Bayard are still in Saint-Nicolas producing carriage clocks, alarm clocks, and a wide range of battery driven wall clocks. Fig 2.

BECKER, Gustav. c 1870-1920.

Fig 3

German movement manufacturer who is known to have produced movements that look similar to the French pendule de Paris movement. Traded as United Clock Factories from 1899 to 1924. Various stamps were employed on Becker products, but they usually included the motif shown. Fig 3.

BOULT. See CRESSON, BOULT.

BOURDIN. A.E., Paris: Rue de la Paix, 1840-1860. Rue Castiglione, 1870.

Bronze Medal, Paris, 1849.

Exhibited in Paris,1867.

BREVETE S.G.D.G.

Frequently found on French movements, this is not a maker's mark but the patent registration mark meaning Breveté sans garantie du Gouvernement (Patented, without Government guarantee of quality).

247

BROCOT, Louis Gabriel, Paris.

Father of the renowned Brocot clockmaking family. His name is usually associated with an adjustable suspension, and a visible escapement that was widely used in French marble clocks. He worked in Paris from 1820 until 1850 (Smith). The mark L. BROCOT with a star underneath is often found stamped on the brass ends of a Brocot suspension spring. These springs have been seen on movements that date from the end of the nineteenth century and one should not assume that they come from the Louis-Gabriel period.

1826   Exhibited a rack striking clock movement (Tardy).
1840   October 9th: Patents granted to last five years 'for improvements to clockwork.' Altogether there were seven variations on an adjustable suspension (Booth).
1840   November 14th: Patent granted for improvements to above (Booth).
1840   June 20th: Patent granted for improvements to the above and 'a new method of facilitating the regulation of pendulums' (Booth).
         See Chapter Five where these improvements are detailed in full.

BROCOT, Achille, Paris, (1817-1874).

Achille was the eldest son of L.G.Brocot. In 1856 he was treasurer to La Société des Horlogers in Paris. Awarded First Class medal at the 1855 Paris exhibition. His trademark was advertised in the 1864 Alamanach Artistique (see Allix and Bonnert) as AB in a star. Fig 4.

Fig 4

BROCOT, Achille et Fils, Paris.

A. Brocot's son succeeded him in the business, to be followed by Gustave Guibaudet (Tardy). See also Guibaudet.

BRITISH UNITED CLOCK COMPANY (BUCC). Birmingham, England, 1855-1909.

The company was formed to manufacture clocks on the American system. Awards: Adelaide 1887. Melbourne & Sydney 1888. Paris 1889. Fig 5.

Fig 5

CHATEAU FRERES & Cie. France.

Their 1908 catalogue showed a black marble clock with Brocot visible escapement and mercurial pendulum priced at 100 Francs.

COMPAGNIE ANONYME DES BRONZES. Belgium: Rue D'Assaut 22, Brussels.

This mark was seen on the dial of a bronze and marble clock garniture sold at Sotheby's, London, December 5th 1985.

CONNELL, William. London, 1817-1862.

Known to have sold French marble clocks. Black marble clock with visible escapement was bought by Connell, circa 1847-1850, for £5 8s and sold for £8 8s.

COUAILLET Frères, Armand (born 1865), Ernest, Gustav, and Henri. Saint-Nicolas-d'Aliermont, 1891-1925.

All the above were brothers and worked in Saint-Nicolas-d'Aliermont. They formed a company in 1891 to produce pendule de Paris movements. Armand trained with VILLON. At the age of 27 he started Couaillet Frères with his brothers and some colleagues. When the company failed around 1930 he worked with Lebrejale. See Chapter Two.

CRESSON, BOULT. Paris: Monmartre 9.

A clock fitted with a PONS movement and a silk suspension had this name and address on the dial. In the Sotheby's, London, catalogue of July 1988, it was dated circa 1840.

CROUTTE, Charles Antoine. Saint-Nicolas-d'Aliermont.

Grandfather of Augustin CROUTTE. Said to be 'Premier horloger de Saint-Nicolas-d'Aliermont'.

CROUTTE, Augustin. Saint-Nicolas-d'Aliermont.

Grandson of Charles Antoine CROUTTE. Founded AUGUSTIN CROUTTE & Cie. in 1847. His report on its foundation, ('Etablissment d'une Fabrique d'Horlogerie'), anticipated the production of 10,000 pendule de Paris movements in 1847. Sizes and quantities were:- 2¾ pouce 2,000; 3 pouce 4,000; 3¼ pouce 2,000; 3½ pouce 500; 3¾ pouce 500; 4 pouce 1,000. (For a definition of pouce see Chapter Four).

CROUTTE, Mathieu. Saint-Nicolas-d'Aliermont.

Worked with Honoré PONS (q.v.) from c 1807 onwards.

DZ. See DELLETREZ.

DAY. London.

In 1807, he made marble clock cases for the English clockmaker Vulliamy.

DELAUNE, Monsieur et Madame. Saint-Nicolas-d'Aliermont.

Governors of the the unsuccessful School of Horology that was founded at Blesdel near Saint-Nicolas.

DELEPINE & CAUCHY. Saint-Nicolas-d'Aliermont.

Believed to be successors to PONS (q.v.), who continued to use the PONS mark. Gold medal, Paris 1900.

DELETREZ, Jean Baptiste. Paris: Rue de Berry au Marais, 1850; Rue Charlot, 1870. London, 1863-1875 (wholesaler).

Awarded First Class medal at Paris Exhibition of 1855. The mark usually appears as the letters 'Dz'. Tardy notes a connection with A. Brocot.

DENIS Frères. Saint-Nicolas-d'Aliermont: 68. Rue Edouard-Cannevel.

Formed by Gustav-Louis Denis in 1874, the company specialised in cutting wheels and pinions for the movement makers of the town. After the 1914-18 war, the two sons Ernest and Georges joined the company. The company was still in existence in 1982.

DESBOIS. London, 1844-1875.

Known to have sold French marble clocks. One striking clock was sold in 1855 for £4 14s.

DESSIAUX. Saint-Nicolas-d'Aliermont.

Mentioned by L'Abbee Decorde (see Bibliography) in 1877 as being an active horologist. Movement maker, 1889 (Tardy).

DULOT/DUSOT. Saint-Nicolas-d'Aliermont.

Mentioned by L'Abbee Decorde in 1877 (see Bibliography) as being an active horologist.

DUVERDY & BLOQUEL.

See Bayard.

DUVOYE. Paris.

Registered a patent in 1852 for a type of clock suspension. The mark DUVOYE appears on the adjusting wheel of some suspensions. (See page 89).

ELKINGTON & Co. Birmingham, England.

Fig 6

This mark appeared on the dial of a French marble clock fitted with a Marti movement made c 1880. Elkington was probably a wholesaler.

EXACTA. Saint-Nicolas-d'Aliermont.

Fig 6. See LEBREJALE.

FABRIQUE D'HORLOGERIE DE SAINT-NICOLAS. Saint-Nicolas-d'Aliermont.

Founded sometime after 1807 by Honoré PONS (q.v.) as a form of guild or co-operative for local clockmakers. In 1819, Paris, awarded two silver medals for clock movements. Award made under title of Exposition Collective de Saint-Nicolas-d'Ailermont.

FARCOT, E. Paris, 1858; Rue des Trois Bornes from 1860 to 1890.

In 1859 patented a visible escapement. Sothebys of London sold (17th Dec, 1987) an alabaster clock with a twin-wheeled visible escapement. The movement was marked FARCOT. See also Tardy.

251

FRODSHAM, Charles. London.

The English company of Charles Frodsham imported many carriage and marble clocks during the period 1850-1914. A white marble clock, with movement number 18205 and a Brocot calendar, was presented as a confirmation present to Prince Alfred, son of Queen Victoria, on April 5th 1860. Some Frodsham marble clocks were marked Charles Frodsham - Paris, although the company did not have an office in that city. See ARNOLD, J R.

FURTWANGLER, Lorenz. Germany.

Maker of German movements that are similar to French pendule de Paris movements. Founded a clock factory in 1836 at the age of 29. Became Furtwangler Söhne in 1868. Clocks are said to have been of the finest materials and craftsmanship. The initials L.F.S. frequently appear in the trademarks.

GASTON JOLLY. (Father) Paris: Rue de Pave-Saint- Sauveur, 1806; Boulevard Poissonniére, 1815-20.

Gaston Jolly, (Son). Rue du Pave-Saint-Sauveur, 1812-20.

GIBAUDET, G. E. France.

Successor around 1889 to the company formed by the Brocot family.

G.L.T. See GUILMET.

GUIGNON. Saint-Nicolas-d'Aliermont.

Mentioned by L'Abbee Decorde in 1877 (see Bibliography), as being an active horologist.

GUILMET, A R.

Exhibited a mystery clock in Paris, 1867, with the mark G.L.T. (Tardy).

HAAS & Söhne. Germany. Founded in 1867 by Philip Haas and dissolved in 1920.

Movement manufacturer who is known to have produced movements that look similar to French pendule de Paris movements. Some marks include the initials P.H.S. For more details see Bibliography (K. Kochman).

H.A.C. In English: **Hamburg-American Clock Company.**

H.A.U. In German: **Hamburg-Amerikanische Uhrenfabrick.**

Fig 7

German movement manufacturer known to have produced movements that look similar to French pendule de Paris movements. Founded by Landenberger & Lang in 1879. Registered as HAU in 1886. In 1891 a crossed arrows motif was registered; in 1893 a Beehive trademark; and in 1905 a Lamp of Wisdom trademark. HAU was eventually bought out in 1930 by their greatest competitor, JUNGHANS (q.v.). The mark below was registered in Hamburg on 18th February 1892. See HIRST Bros. Fig 7.

HENSON, Robert. London, 1851-1863.

Maker of marble cases for English clocks.

HENSON, Samuel, London, 1875-1881.

Maker of marble cases for English clocks.

Fig 8

HIRST Bros. Manchester.

Catalogue (see Bibilography) circa 1910 shows 26 marble clocks fitted with either H.A.C movements (q.v.) or French movements. Fig 8. See Chapter Two.

HOLLINGUE or HOLINGUE. Saint-Nicolas-d'Ailermont.

Mentioned by l'Abbee Decorde in 1877 (see Bibliography), as being an active horologist. Bronze Medal for roulants, Paris 1849. (See also Allix and Bonnert).

HOWELL & JAMES. London, 1820 to c 1862.

Several of the company's marks have been seen on the dials of French marble clocks. After 1840 the name Howell James & Co, and between 1867 and 1889 the name Howell James & Co, London and Paris, was used on the dial. The mark Howell and James to the Queen London appears to refer to Queen Victoria. The company

253

is known to have had large premises at 5, 7 and 9 Regent Street, London, from 1820 until 1962 at least.

JAPY, Frédéric. Beacourt, France. Born 22nd May, 1749; died 23rd June, 1812. Also at Badeval, France; 108 Rue du Temple, Paris; and Le Quai Jemmapes, Paris.

Apprenticed to Jean-Jacques Perrelet in Le Locle, Switzerland, he founded the first workshop in Beaucourt in either 1770 or 1772 to produce ebauches (unfinished watch movements). Married Catherine Suzanne on 16th February 1773 and moved to Montbeliard region in 1774, returning in 1777 to Beaucourt and employing 50 workers. By 1801 there were 300 workers producing 100,000 ebauches a year. Frédéric Japy was regarded as the 'Father of the French horological industry of the 19th century.' His sons, who joined the business, were Frédéric Guillame known as Fritz (1774-1834), Louis Frédéric (1777-1852), Jean-Pierre (1785-1863), Charles (1792-1821), and Frédéric known as Fidot. The first three sons worked at Beaucourt while the latter two were responsible for Badeval. Between 1810 and 1888 the company manufactured over 6 million blank clock movements.

1804   Received Croix d'Honneur from Napoleon.
1806   Japy sons taken into partnership.
1810   Converted a water mill at Badeval, near Beaucourt, to provide clock movement factory, that subsequently produced most of the Japy movements seen in French marble clocks.
1812   23rd June. Death of Frédéric Japy (senior).
1821   Ingenu Japy, son of Fritz, became manager of Badeval factory.
1821   February 3rd. Formation of Japy Frères, incorporating Fritz, Louis and Pierre Japy.
1837   1st August. Name of Japy Frères re-registered. Partners of the company in 1837 were: Charles Louis Meiner, Auguste Julien Japy, Ingenu Japy, Charles Adolphe Japy, Louis Octave Japy, and Louis Auguste Monnin.
1850   Production started of watch cases to be fitted with Japy watch movements.
1854   Death of Fritz Japy. Japy Frères dissolved.

Fig 9

Fig 10

Fig 11

Fig 12

Fig 13

TROCADERO

Fig 14

Fig 15

1854   1st August. Formation of Société Japy Frères et Cie. Name re-registered in 1873, 1882, and 1910.

1860   Production starts of clock cases to be fitted with Japy clock movements.

1878   Adolphe Japy, Beaucourt. managing director of Société Japy Frères et Cie.

1882   Start of marble working.

1928   28th February. Formation of Etablissments Japy Frères with effect from 18th May 1928 until 2000.

1933   Rationalisation of Japy Company starts.

1936   Clock production ends at Badeval.

All the awards were for clock or watchmaking; items 1867 to 1879 are believed also to be for horology but this cannot be verified. The text of the 1855 award reads: 'Large scale mechanised manufacture of watch ebauches and blanc roulants for clocks. Good quality and low price' Chavigny and Boulay (see Bibliography) make the following comments about Japy:

'The early movements, fitted with a silk suspension, carried the marks of: JAPY - Médaille d'Or 1823-27-34-39-44-49.

After 1850 until 1858 the movements were marked: JAPY FRERES - Médaille d'Or. This was replaced by: JAPY FRERES - Grande Médaille d'Honneur 1855. Finally around 1888 came the mark: JAPY FRERES & Cie - Médaille d'Honneur'.

This particular mark has a cross in the middle that is similar to that of the Legion d'Honneur, Fig 9. The Japy pendule de Paris movements usually had the medallion type of stamp mentioned above. Cheaper tic-tac movements were also fitted to marble clocks. Japy tic-tac movements have been seen with marks similar to some of the following. (See Chapter Two).

Fig 10.   Japy Frères et Cie. Beaucourt. Reg. 18th February 1884, Leipzig, Germany.

Fig 11.   Japy Frères et Cie. La Chaux de Fonds. Reg. September 1st 1887, Leipzig, Germany.

Fig 12.   Société Japy Frères, Beaucourt. Reg. August 9th 1898.

Fig 13.   Société Japy Frères, Belfort. Reg. 16th March 1905.

Fig 14.   Société Japy Frères, Belfort, 1st May 1908.

Fig 15.   Société Japy Frères, Belfort, Reg 29th December 1913.

Gold Medals:	Louis XV111	1819
	ditto	1823
	Charles X	1827
	Louis Philippe	1834
	ditto	1839
	ditto	1844
	ditto	1849
Grand Medal	Great Exhibition, London	1851
1st Class Medal	New York Exhibition	1853
Grand Medal of Honour	Exp. Universelle, Paris	1835
ditto		1855
ditto		1867
Grand Diploma of Merit	Exp. Universelle,Vienna	1873
Grd. Prix d'Honneur	Exp. Universelle,Paris	1879
Grd. Diplôme d'Honneur	Exp. Regionale,Besançon	1879
ditto	Exp speciale d'horlogerie de La Chaux-de-Fond	1880

JAPY et Cie, L.P. Berne, near Seloncourt. Doubs.

Louis P was son of Louis Japy and grandson of Frédéric Japy. Second class medal at Paris Exhibition of 1855 awarded to 'Japy, Fils, L'. A movement marked LEP Japy et Cie - Médailles, c 1870, has been attributed to Louis Japy.

JAPY, MARTI, ROUX. Paris: Boulevard du Prince Eugene, 1870.

These three manufacturers from the same area of France formed a company for the sale of their joint goods through a Parisian depot. Emile Japy manager in 1860. Exhibited at Paris, 1867.

JAPY, MARTI, ROUX, MOUGIN.

Allix and Bonnert note that the name Mougin appeared on the wrapper of some blanc roulants, along with the names Japy, Marti, and Roux. Address in 1880 believed to be 75 Rue Turrene, Paris. See JAPY, MARTI, ROUX. Also MOUGIN.

JUNGHANS. Black Forest, Germany. c 1870.

German movement manufacturer known to have

Fig 16

produced movements that look similar to the French pendule de Paris. A frequently employed motif consisted of a star containing the word Junghans. In 1888 the star had five points and in 1890 eight points. Fig 16.

For more details see 'The Junghans Story' (Antique Clocks Publishing, Concord, California.) See Bibliography (K. Kochman).

L.H.S. See FURTWANGLER.

LEMAITRE & BERGMANN. Paris, 1848. London, 1857.

Became C.A. Richard et Cie. See RICHARD.

LENZKIRCH. Lenzkirch, Germany. c 1875.

German movement manufacturer known to have produced movements that look similar to the French pendule de Paris movement.

LEBREJALE. Saint-Nicolas-d'Aliermont. Rue de Franche Comte, Paris.

Lebrejale traded as EXACTA. Founded in 1807 and still going in 1933, when a catalogue showed two complete pendule de Paris movements. Armand COUAILLET worked with Lebrejale c 1930. Fig 6. See Chapter Two.

MARC, Henry. Paris.

Known vendor of clocks fitted with pendule de Paris movements. Japy carriage clocks have been seen with the stamp Hry. MARC PARIS on them. (Allix and Bonnert).

MARTI & Cie, S. France.

Fig 17

This name appears frequently on the movements of French marble clocks. Unfortunately little is known of the company. Second Class Medal at Paris Exhibition of 1855. Known markings are SAMUEL MARTI - Médaille de Bronze, awarded in 1860. Also Médaille d'Or, Paris, 1900. Fig 17.
Loomes refers to J MARTI & Cie, France, c 1860. Fritz MARTI had premises in Vieux-Charmont in 1867 and exhibited blanc roulants in Paris Exhibitions of 1889 and

257

1900. Fritz MARTI Et Fils awarded a Médaille d'Or at Paris in 1900.

MARTIN, Emile J. Saint-Nicolas-d'Aliermont.

Mentioned by L'Abbee Decorde in 1877 (see Bibliography) as being an active horologist. Exhibited movements at Rouen in 1856 and at Paris in 1867.

MARTIN et SAUTEUR FRERES. Saint-Nicolas-d'Aliermont.

Exhibited at Paris in 1867.

MCCABE, James. Royal Exchange, London, 1781-1883.

Known to have produced a black marble clock fitted with an English fusee movement.

MEGNIN. Saint-Nicolas-d'Aliermont.

Mentioned by L'Abbee Decorde in 1877 (see Bibliography) as being an active horologist.

MIROY FRERES. Paris.

A white marble and ormolou clock from c 1860 has been seen with a VINCENTI movement inscribed Miroy Fres. Paris.

MOSER, Georges. Paris: Rue du Grenier, Saint-Lazare, 1830. Boulevard du Temple, 1840.

After 1860 the mark became MOSER et Cie. Exhibited in London, 1851.

MOSER et MARTI.

Tardy suggests that there is a connection with MARTI (q.v.) of Montbeliard and offers the following dates: Montbeliard, 1838, and Paris, Route d'Orleans, 1840-1850. MOSER is reported to have started a clock movement factory in Montbeliard in 1834.

MOUGIN, A.D.

Believed to be the Mougin of Hermincourt who exhibited roulants in 1900. Name also associated with JAPY, MARTI, ROUX, and MOUGIN. Fig 18.

Fig 18

258

MOYNET, M.M. et Cie. Paris.

Tripplin (see Bibliography) noted in 1889 that this company of watch and clock material dealers commanded the respect of all French watch and clockmakers.

PAYNE. London: 163 New Bond Street.

PAYNE & Co. London: 13 New Bond Street, 1825-1881.

Became PAYNE & Co. after 1851. Large importer of complete French marble clocks and of French movements.

PONS, Honoré, Saint-Nicolas-d'Aliermont.

Also known as Pons de Paul. Died c 1857 (Will was read on 19th July 1857). Sent to Saint-Nicolas-d'Aliermont in 1807 by the French Minister for Clocks and Watches, M. Champigny, to boost the local horological industry. Aided in his work by Mathieu CROUTTE (q.v.). Pons founded a guild, or co-operative, called FABRIQUE D'HORLOGERIE DE SAINT-NICOLAS (q.v.). His own company produced chronometers and pendule de Paris movements marked Honoré Pons à Paris.

His awards were:-

1823 Paris Silver Medal for mechanisms.
1827 Paris. Exhibited four new escapements at the Louvre.
1839 Paris Gold Medal.

The name PONS was used by his successors DELEPINE & CAUCHY. Fig 19.

Fig 19

R.C. See RICHARD et Cie.

RAINGO FRERES. Paris: Rue Vielle du Temple 1829 -1900.

Founded by Z. Raingo about the beginning of the 19th century.

RICHARD et Cie, C.A. Paris: Rue de Bondy 32 from 1848. London: 24 Cannon Street from 1867.

This mark was seen on the back plate of a Japy Frères

movement from after 1888. The case was green onyx. Allix and Bonnert note that the company was originally founded in Paris as Lemaitre & Bergmann 1848, and a branch opened in London in 1857. Fig 20.

Fig 20

ROLLIN. Paris: Rue de la Cordèrie, 1840-1850. Rue de Bretagne, 1860-1870.

Successor to DOMANGE. A white marble and ormolou clock marked Rollin a Paris was sold at Sotheby's, London, in June 1985. The catalogue indicated c 1860.

ROSETTI, Paul. Saint-Nicolas-d'Aliermont.

An Italian horlogist who worked with HONORE PONS.

ROUX. Paris: Boulevard du Prince Eugene, 3. Montbéliard, France.

Roux succeeded VINCENTI (q.v.) in Montbéliard and was known to be manufacturing roulants in 1867. Silver Medal 1849. Exhibited roulants, Paris, 1851. See also MARTI, and JAPY.

THIEBLE.

Inventor of a pendulum modification used in French marble clocks and attributed to the year 1865. (See Chapter Three).

SAUTEUR. Saint-Nicolas-d'Aliermont.

Mentioned by L'Abbee Decorde in 1877. See Bibliography, as being an active horologist.

SEARS ROEBUCK.

American mail order distributor of Waterbury clocks.

SEIGNEUR. Saint-Nicolas-d'Aliermont.

Mentioned by L'Abbee Decorde in 1877 as being an active horologist.

S.G.D.G. See Breveté.

UNITED CLOCK FACTORIES. See Becker, Gustav.

VALLET.

A type of pendulum suspension has been attributed to Vallet. See Chapter Four.

VENOT FRERES et Cie. Paris: 134 Rue du Temple.

Wholesalers of Japy products.

VILLON, Albert. Saint-Nicolas-d'Aliermont.

Founded a factory in 1867 that he ran until 1887. In 1887 Paul Duverdy became manager and ran the company until 1910 when Villon died. Duverdy was joined by Joseph Bloquel who was manager until after 1922. The company during this period was presumably called Duverdy & Bloquel. See also BAYARD. In 1898 the company was the largest in Saint-Nicolas with an annual production of 20,000 carriage clocks, 20,000 mantle clock movements and 60,000 alarms. (See Chapter Two). Villon is believed to have been the driving force behind the unsuccessful School of Horology that was founded at Blesdel near Saint-Nicolas.

VINCENTI. Montbéliard.

Fig 21

One of the largest movement makers next to JAPY. Croutte (see Bibilography) stated in 1847 that Japy and Vincenti were between them producing 86,000 blanc roulants a year. Vincenti is believed to have been a Corsican who started his factory in Montbéliard c 1823. He used machines that he had invented to produce rough movements for clocks and watches. The business failed in 1824 and the factory from there onwards was run by ROUX. In 1825 between 40 and 50 workers were employed (Allix and Bonnert). Production of movements is believed to have ceased in 1870. Known marks are: VINCENTI - Médaille d'Argent 1834. VINCENTI & Cie - Médaille d'Argent 1855. Fig 21.

VULLIAMY, Benjamin. London, c 1781-1820. Benjamin Lewis, London, c 1809-1854.

Makers of exceptionally fine marble clocks fitted with English movements.

261

WATERBURY CLOCK CO. Connecticut, U.S.A. Established 1857.

Producers of imitation French marble clocks. Established originally in 1812 as The Benedict and Burnham Manufacturing Co. to produce brass. The Waterbury Clock Co. was established in 1857 and by 1873 had grown sufficently large to require larger premises. In 1917, 3,000 workers were employed by the company. As with their French counterparts, Waterbury's business declined after the First World War.

WERNER, C. c 1880-1910.

German movement manufacturer who is known to have produced movements that look similar to the French pendule de Paris movement. Trademarks sometimes employed an inverted horseshoe motif or simply the name Verna.

Author's note

Many kind enthusiasts provided me with valuable information, some that they previously considered to be unworthy of attention. I am still collecting material and would be grateful for any help. Please send correspondence care of the publishers. All letters will be acknowledged by the author, who will return promptly any photographs, drawings, and documents after inspection.

Nicolas M Thorpe.

BIBLIOGRAPHY

Two bibliographies are listed. One is a suggested selection of books for a newcomer to French marble clocks. The other gives the reference sources used for the material contained in this book.

SELECTED BIBLIOGRAPHY

1   *Allix and Bonnert.*
Carriage Clocks - Their history and development.
Antique Collectors Club, Woodbridge, Suffolk, England. 1981. Many makers of carriage clocks also produced the pendule de Paris movements that were used in the French marble clock. This book gives a well-illustrated and informative history of many makers, such as Japy and the companies from Saint-Nicolas d'Ailermont.

2   *Brewer.*
The Country Life Collector's Pocket Book of Clocks.
Hamlyn, Northants, England. 1983. An inexpensive and compact book that covers most of the domestic clocks that will be encountered by the enthusiast. There are a few pages that deal affectionately with the French marble clock. Contains more useful information than some of its grander contemporaries.

3   *Gazeley*
Watch and Clockmaking. Out of print in 1986. Unfortunately this book is out of print; it is one of the few that deals well with the subject of restoring a French movement of the type found in marble clocks. Usually available in public libraries. Donald de Carle's book (see below) is a possible substitute.

4   *Shenton, A and R.*
The Price Guide to Clocks, 1840-1940. Antique Collectors Club, Suffolk, Woodbridge, England. 1985. Chapter One deals exclusively with French marble clocks. It provided the sole source of information for many years and was the only book mentioned by the British Museum when I wrote to them in the early days of researching this book.

5 *Tardy.*
Dictionnaire des Horlogers Français. Paris. 1971. The standard reference work for anyone wishing to date French clocks. Unfortunately it is very expensive although only available in paperbinding.

GENERAL BIBLIOGRAPHY

The publications listed either provided reference material for this book, or have material that relates to French clocks. Journals and catalogues are listed first, then books and papers.

Journals and Catalogues

6 *Antiquarian Horology.*
Journal of the Antiquarian Horological Society, Ticehurst, England. Quarterly.

7 *Clocks.*
Journal published in Hemel Hempstead, England. Monthly

8 *Horlogerie Ancienne.*
Review of the Association Française des Amateurs d'Horlogerie Ancienne. Besançon, France. Quarterly.

9 *English Clock Factory, The.*
H. Williamson Ltd catalogue. Salisbury, England, c 1900.

10 *Egypt.*
The Inspiration of Egypt. Catalogue of exhibition at Brighton Art Gallery, Brighton, England. 1983.

11 *Great Exhibition of 1851.*
Catalogue. London. 1851.

12 *Hirst Bros and Co Catalogue (c 1910).*
(Prickett). Reprinted Maxwells, Birmingham, 1982.

13 *Japy Frères*
Catalogue 1898. Beaucourt, France, 1896.

14 *Paris Exposition Universelle de 1855.*
Rapports du Jury International: Grande Médailles d'Honneur et Médailles d'Honneur. Paris. 1855.

15 *Paris Universal Exhibition.*
Illustrated London News, London. 1878.

16 *Paris Exhibition of 1900.*
Report by Roger Marx, England. 1901.

17 *Restoration, Studies in.*
Victoria and Albert Museum, South Kensington, London.

18 *St Nicolas.*
Histoires et Petites Histoires. Saint-Nicolas-d'Aliermont, 1982.

19 *Sotheby Saleroom Clock Catalogues.*
Sotheby and Co, London, England.

20 *Waterbury Clock Company Catalogue.*
No.131. American Clock and Watch Museum, Connecticut, USA.

Books and Articles

21 *Baillie.*
Watch and Clockmakers of the World Vol I. NAG Press, Colchester, England. (See *Loomes* for Vol II)

22 *Barker.*
The Arthur Negus Guide to English Clocks. Hamlyn, Feltham, England. 1980.

23 *Berger.*
La Calcaire Devonien: Le Petit Granit et les Pierres de la Meuse. Belgium. 1890.

24 *Berton and Javel.*
Exploitation en Carrières et Transformation en Usines des Marbres et Pierres Calcaires Utilises pour la Construction et l'Ornamentation. Belgium. 1970.

25 *Blagrove.*
Marble Decoration and the Terminology of British and Foreign Marbles. Crosby, Lockwood and Sons, London. 1888.

26 *Booth.*
New and Complete Clock and Watchmakers Manual. New York. 1877.

27 Booth.
The Marble Workers Manual. Philidelphia, USA. 1886.

28 Britten.
Horological Hints and Helps. First published 1929. Baron, Suffolk, England. Reprint 1977.

29 Britten.
Old Clocks and Watches, and their Makers. Methuen/Spon, London 1982.

30 Britten.
Watch and Clockmakers' Handbook, Dictionary and Guide. Eyre, Methuen/Spon, London. 1978.

31 Bromley.
Books and Manuscripts in the Library of the Worshipful Company of Clockmakers. Sotheby, Parke Bernet, London, 1977.

32 Cauchy.
Memoire sur la Constitution Geologique de la Province de Namur, Brussels, 1825.

33 Cemeraman.
Les Roches Calcaires de la Belgique. Belgium. 1947.

34 Chapuis.
De Horlogis in Arte. 1954.

35 Chavigny and Boulay.
La Pendule de Paris 1750-1950: Sa Description, sa Datation, sa Réparation. France. 1983.

36 Cobban.
A History of Modern France, Vol 2 1799-1945. Penguin Books, London. 1961.

37 Crouette.
Etablissment d'une Fabrique d'Horlogerie a Saint-Nicolas-d'Aliermont. Saint-Nicolas-d'Aliermont, France. 1847.

38 Daniels.
Watchmaking. Sotheby and Co, London. 1981.

39   Davreux.
Essai sur la Constitution Geognostique de la Province de Liege. Brussels. 1833.

40   De Carle.
Practical Clock Repairing. NAG Press. Colchester. 1983.

41   De Carle.
The Watchmaker's and Model Engineer's Lathe. Hale. London. 1985.

42   Decord.
Histoire d'Ailermont. 1877.

43   Demaret.
Gisements, Exploitation Travail et Commerce des Marbres en Belgique et à la Entranger. Belgium. 1866.

44   Drummond.
The Evolution of Clockwork. England. 1972.

45   Dumon.
Quelques Observations sur le Marbre Noir de Mazy. Belgium. 1933.

46   Dumon.
Sur le Marbre Noir de Mazy. Belgium. 1934.

47   Fleet.
Clocks. London. 1961.

48   Ford.
The Black Marble of Ashford-in-the-Water, Derbyshire. The Liverpool and Manchester Geological Journal, Vol 2, Part 1. England. 1958.

49   Ford and Reinwerts (Ed).
 Lead Mining in the Peak District. England.

50   Gazeley.
Clock and Watch Escapements. Newnes, London, 1956.

51   Gilbert.
La Marbrèrie: Choix de Dessins Representant des Travaux de Marbrèrie etc. Paris. 1866.

52 *Gloag.*
Victorian Taste: Some Social Aspects of Architecture and Industrial Design 1820-1900. David and Charles, Newton Abbot, England. 1962.

53 *Grant.*
The Marbles and Granites of the World. Grant, London. 1955.

54 *Groessens.*
L'Industrie du Marbre en Belgique. Service Geologique de Belgique. Brussels. 1981.

55 *Gros.*
Echappements d'Horloges de Montres. Paris. 1922.

56 *Harris.*
Industrial Archaeology of the Peak District. David and Charles, Newton Abbot, England. 1983.

57 *Hasluck.*
The Clock Jobber's Handbook. London. 1914.

58 *Illustrated Journal for Mechanics.*
How to make a Marble Clock Case. England. February 29, 1896.

59 *Jagger.*
Royal Clocks: The British Monarchy and its Timekeepers. England. 1983.

60 *Jendritzki and Matthew.*
Repairing Antique Pendulum Clocks. Lausanne. Switzerland.

61 *Johnson.*
The Antiques Collecting Directory. Lists of restorers, dealers, specialist suppliers etc. Pan, London. 1983.

62 *Kaisin.*
Observations sur le Marbre Noir de Golzinne. Belgium. 1912.

63 *Kemp.*
The Victorian Revival in Interior Design. Columbus Books, England. 1986.

64 Kochman.
Clock and Watch Trademark Index. European origin. LaRose, Greensboro, USA. 1978.

65 Laborde.
De L'Union des Artes et de L'Industrie. Two volumes. Impr_emiere Imperiale. Paris, 1856.

66 Landes.
Revolution in Time: Clocks and the Making of the Modern World. Harvard University Press, USA. 1983.

67 Loomes.
Watchmakers and Clockmakers of the World, Vol II. (See also *Baillie*). NAG Press, Colchester, England.

68 Le Mausolee (magazine in French).
Quelques Carrières de Marbres Noirs et Rouges Belges. Belgium. 1973.

69 Lye.
Mineral and Rocks. Kingfisher Books, London. 1979.

70 Nicholls.
English Bracket and Mantle Clocks. Blandford Press, Poole, England. 1981.

71 Planchon.
La Pendule de Paris: Son Evolution Artistique etc. Paris. 1921.

72 Renwick.
Marble and Marble Working: A Handbook for Architects, Sculptors, Quarry Owners and Workers, and all engaged in the decorative Industries. London. 1909.

73 Saunier.
Treatise on Modern Horlogy. Foyles, London. 1975, Reprint of 1861 edition.

74 Smith, A (Ed).
The International Dictionary of Clocks. Hamlyn, Northants, England 1979.

75 Smith, E.
Clocks, Repair and Maintenance. England. 1977.

76 *Smith, E.*
Repairing Antique Clocks. England. 1975.

77 *Smith, E.*
Striking and Chiming Clocks: Their Working and Repair.
David and Charles, Newton Abbot, England. 1985.

78 *Spon.*
Workshop Receipts: For the use of Manufacturers,
Mechanics and Scientific Amateurs'. Spon, London. 1895.

79 *Tardy.*
French Clocks the World Over. Three volumes. Paris.
1981.

80 *Tardy.*
La Pendule Française. Paris, 1970.

81 *Tripplin.*
Watch and Clockmaking at the Paris Exhibition of 1889:
Being an Account and Comparison of the Exhibits in the
Horological section of the French International
Exhibition. London 1889.

82 *Trawinski.*
Manuel D'Archeologie Grecque et Romaine, Vol I - La
Grece. Rothschild, Paris. 1902.

83 *Tyler.*
American Clocks for the Collector. Hale, London. 1981.

84 *Vinter.*
Histoire des Establissments Japy Frères (1777-1943).
Japy Frères, Beaucourt, France. 1944.

85 *Ward.*
The World and Its Workshops: Guide to the Great
Exhibition. London. 1851.

86 *Wilding.*
How to Repair Antique Clocks. Three Vols. Ashford,
England.

87 *Whiten.*
Repairing Old Clocks and Watches. NAG Press,
Colchester. 1979.

Index